AF699335

Predicting and Measuring Fugitive Dust

Predicting AND Measuring FUGITIVE DUST

Howard E. Hesketh, Ph.D.
Mohammad S. El-Shoboksy, Ph.D.

TECHNOMIC
PUBLISHING CO., INC.

TD 884.5 .H476 1985
Hesketh, Howard E.
Predicting and measuring
fugitive dust

Technomic Publishing Company, Inc.
851 New Holland Avenue
Box 3535
Lancaster, Pennsylvania 17604

©1985 by Technomic Publishing Company, Inc.
All rights reserved

No part of this publication may be reproduced, stored in a retrieval system, or transmitted, in any form or by any means, electronic, mechanical, photocopying, recording, or otherwise, without the prior written permission of the publisher.

Printed in the United States of America

10 9 8 7 6 5 4 3 2 1

Main entry under title:
Predicting and Measuring Fugitive Dust

A Technomic Publishing Company book
Bibliography: p.
Includes index p. 129

Library of Congress Card No. 84-51877
ISBN No. 87762-375-9

Contents

Chapter 2 / Sampling and Monitoring

Chapter 3 / Data Processing

Preface

This handbook provides basic information in a practical manner for studying fugitive particulate matter. It supplements the companion book in this series entitled *Fugitive Emissions and Controls* by presenting detailed procedures for measuring source emissions, calibrating measurement systems, and by suggesting techniques for calculating and presenting results. Sample validation and quality assurance to provide reliable results are emphasized. Step-by-step procedures are included for equipment set up, laboratory and field calibrations, field testing, and laboratory analysis.

The Figures, Tables, and formulas presented here are directly applicable to fugitive dust work as portions of this handbook replace many of the instrument manual instructions and calibration calculations. Complicated theoretical approaches are avoided and practical assistance in the form of data sheets, precautions, advice, and recommendations are included.

The word fugitive suggests that it is not a simple task to obtain reliable source emission data. By understanding the behavior of the fugitive dust, it becomes easier to accomplish this task. In addition, information on the large (respirable) and small (inhalable) fractions may be needed and these discussions are included in this book. We hope that our compilations of data equipment suggestions, techniques, and procedures are useful to you the readers.

HOWARD E. HESKETH
MOHAMMAD S. EL-SHOBOKSHY

Acknowledgements

We sincerely appreciate the help provided by Southern Illinois University Department of Thermal and Environmental Engineering for providing both the resources and assistance used to conduct some of these fugitive studies and for secretarial assistance in preparing this manuscript. In addition, we thank Joyce S. Hesketh for her help in proofreading and Michael A. Ebersohl for all of the drafting in this book. Thanks also go to the contributors of equipment drawings, operating procedures, and photographs used in this book and to David J. Kolaz and Saleem Salaymeh of the Illinois EPA for providing resource materials.

1/Fugitive Dust Behavior

1.1/Introduction

A companion book in "The Environment and Energy Handbook Series" defines fugitive emissions as "those pollutants which are generated from open sources exposed to the air and are not discharged into the atmosphere in a confined flowstream" [1]. It further defines fugitive *emissions* as industry-related and fugitive *dust* as all other and then assumes both to be synonymous (gases per se are not discussed). This book will adhere to these definitions.

These materials enter the ambient air and either become part of it or settle from it onto some surface. During the period they are part of the atmosphere they become, to some degree, subject to the effects of the fluid (air) in which they are suspended. On a megascale, the earth's mantle of air is a fluid at rest and follows the law of hydrostatis. On a mesoscale, the air also moves to transport the particles and this is discussed in Section 1.2.

In addition to these effects, these emissions do not remain as emitted and may change chemically and/or physically. As an example of matter in the atmosphere:

1. Gases can react with other gases to form secondary particles
2. Gases can react with, condense on, or become adsorbed onto existing particles
3. Particles can evaporate and/or volatize and thereby become gaseous in form
4. Particles may agglomerate due to Brownian motion which causes them to collide and merge together.

These actions take time and depend on the concentration. Hesketh [2] shows that Brownian motion coalescence is described by a second order equation which yields

$$N_2 = [Kt + 1/N_1]^{-1} \tag{1}$$

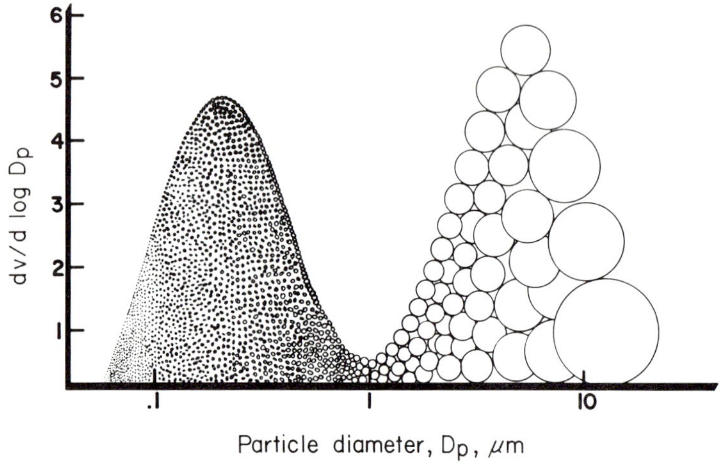

Figure 1.1. Distribution of atmospheric aerosol (Courtesy TSI, Inc. [3]).

where N_1 and N_2 are initial and final particle concentrations per cm^3 of air, t is time in seconds and K is a constant ranging from 0.51×10^{-10} to 3.7×10^{-10} for "larger" and "smaller" particles respectively.

The actions described above plus gravitational settling results in a typical size distribution of particles in the normal atmosphere such as shown in Figure 1.1. This is an artist's version of a mass frequency plot of atmospheric aerosols depicted by number and size. (For our purposes the function [dv/(d log D_p)] represents frequency.) This figure shows a bimodal distribution, although trimodal also occurs. The fine particles, which are shown here as $<1 - 2$ μm diameter, are normally the result of totally suspended particles (TSP) which may have been transported by the atmosphere long distances (e.g. 100 miles). They could have resulted from direct emissions of fine particles and from chemical and/or physical interactions as noted. The large particles are those which have been released from a relatively nearby source. They will settle from the atmosphere. Note that by number there would be about 10^4 to 10^5 more fine particles than course particles.

The short and long term interactions are present and can result in different effects for the same atmospheric particles. The reader is cautioned that this monograph concentrates on the short term behavior and effects of fugitive dust emissions. The emissions and the measuring devices are therefore physically in close proximity (e.g. 15 feet to 1 mile). In these cases the long term influences are usually considered insignificant, depending on the distances involved. For cases where both short and long term influences are important they must both be included.

1.2/Atmospheric Transport and Particle Properties

The Introduction Section 1.1 notes that the resultant ambient particle concentration is a function of both long and short term effects. It also states that this book primarily deals with the short term effects such as occurring within 15 feet to 1 mile from the source. Under these conditions, agglomeration as noted in the Introduction can be neglected except for extremely high concentrations, where moisture condenses or where other external forces are applied which cause the particles to agglomerate rapidly.

The most important particle properties relative to atmospheric transport are size, shape, and size distribution. Some physical properties such as density are very significant while other physical properties and most chemical properties are negligible for short term transport effects. For more detailed discussions on particle behavior see other references such as Hesketh [4].

1.2.1 Size Considerations

Most fugitive particles released into the atmosphere have a log normal size distribution. As such, the sample would plot as a straight line on a log of diameter versus cumulative percent less than stated diameter on the probability axis as in Figure 1.1. If the sample consists of two or more separate products, the plot will consist of two or more separate intersecting straight lines. The two different products could be different chemically such as particles of limestone and soil or they could be the same chemical substance but having been subjected to different physical treatment such as a mixture of crushed limestone from a hammer mill and a ball mill.

Example 1.1

Check the size distribution of the Table 1.1 particulate data from a sample of taconite ore fugitive dust.

First, rearrange the data shown in Table 1.1 into percent by mass for each group. Do this by finding the total mass of sample (3.50g), dividing this into each group mass and converting to percent. For the two largest groups this is:

$$\frac{1.750\ \text{g}}{3.50\ \text{g}}\,100 = 50\%$$

Table 1.1. Taconite Ore Dust Data.

Raw Data Given		Calculate
Sample Mass, g	Size, μm	mass %
1.750	>10	50
0.350	8–10	10
0.385	6–8	11
0.245	5–6	7
0.210	3–5	6
0.158	2–3	4.5
0.193	1–2	5.5
0.105	0.5–1	3
0.105	<0.5	3

$$\frac{0.350 \text{ g}}{3.50 \text{ g}} 100 = 10\%$$

Next, relist the distribution as shown in Table 1.2 to give cumulative percent less than stated diameter. A convenient procedure for this is to start with the smallest group and add all sample mass now shown. These data are plotted on the Figure 1.2 log-probability paper. Note that these data show the taconite ore to be bimodal in size distribution. This is because the bulk of material consists of two different iron ore compounds.

The mass mean diameter of a particle sample is the value that occurs at the 50 percent value (d_{50}) on log probability plot such as Figure 1.2. The mass mean diameter of the taconite ore sample is 10 μm from Figure 1.2.

Sample size distribution is readily obtainable from log-probability plots. The standard geometric deviation (σ_g) values equal:

Table 1.2. Taconite Ore Data in Revised Form.

Stated Diameter, μm	Cumulative % less than
10	50
8	40
6	29
5	22
3	16
2	11.5
1	6
0.5	3

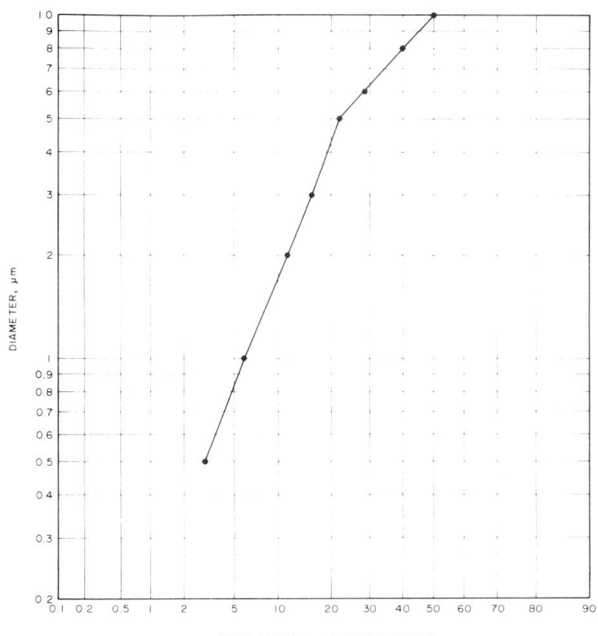

Figure 1.2. Taconite ore dust size distribution.

$$\sigma_g = \frac{d_{50}}{d_{15.87}} = \frac{d_{84.13}}{d_{50}}$$

In the case of bimodal distribution, there can be two values of σ_g.

1.2.2 Settling Velocity

Most of the fugitive particles of significance can be handled by Stokes equations as corrected by the Cunningham factor (C). This means that the Stokes, i.e. sedimentation diameter (d), must be used. The Stokes diameter is defined as the diameter of a sphere having the same density and settling velocity of the particle in question. The average density is used in determination of the Stoke's diameter. Note that the validity of size information based on the average density is dependent on the uniformity of the density. Inhomogeneity of particle composition may be detected by visual inspection under the optical microscope. Note that the Stoke's diameter (d) as defined is for spherical particles.

A particle released in a stagnant atmosphere accelerates and very quickly attains a constant, terminal settling velocity (V_s) of

$$V_s = \frac{d^2 \, g \, \varrho_p \, C}{18 \, \mu_a}$$

where

g = gravitational acceleration (980 cm/sec^2 at sea level)
ϱ_p = particle density
μ_a = density of the air (1.83 $\times$ 10^{-4} g/(cm sec) at standard conditions)

A simplified equation for estimating the Cunningham factor is

$$C = 1 + \frac{(6.21 \times 10^{-4})\, T}{d_p}$$

where

T = temperature in °K (i.e. 273 + °C)
d_p = particle Stoke's diameter in micro meters, μm

As a rule of thumb, C essentially equals 1 for particles larger than 1 μm.

1.2.3 Aerodynamic Diameter

This is defined as the diameter of a sphere of unit density that has the same settling velocity as the particle of concern. The aerodynamic diameter has found popularity because of the interest in aerosol and particulate pollution research was initiated by the public health scientists, environmental scientists, and meterologists. It is usually used in the study of the dispersion of plumes, deposition of particulate matter on surfaces, or in the human respiratory system. The collection efficiency of particulate control equipment is normally specified as a function of the aerodynamic diameter. Most of the sampling instruments classify the particles according to the aerodynamic diameter.

If the aerodynamic diameter (d_a) is obtained from the measuring instruments, then the Stoke's diameter is obtained by

$$d = d_a \sqrt{C_a/(C \, \varrho_p)}$$

where C_a = Cunningham factor with respect to d_a.

1.2.4 Cut Diameter

The cut diameter, d_c is defined as the diameter of particles collected with an efficiency of 50 percent by a collection device. It is convenient to specify this diameter when using inertial impactor type of samplers and it is expressed as a 50 percent aerodynamic cut diameter (d_{50}). The cut diameter depends on the specific design of the sampler instrument and the sampling gas flow rate and temperature. Calibration curves are usually provided with particle fractionating samplers to give the cut diameter associated with the sampling flow rate.

1.2.5 Particle Density

Some properties of particles depend to a great extent on the way by which the particles are generated. Particles may contain air voids due to the nature of their production process or often undergo surface reactions that create inhomogeneity in the original material. The particle density is a property that is largely affected by these circumstances and lead to an apparent density that differs from either the bulk density or the true density. The apparent density is the weight of a unit volume of packed particles. Particles may also appear as loosely grouped agglomerates and may have much lower masses than would be expected with the diameters measured under the microscope. The apparent density (a) is related to the bulk density (b) in terms of the void fraction (v) due to air packets or air space between agglomerated particles as follows:

$$a = b\,(1 - V)$$

The value of V can only be determined experimentally. For accurate determination of fluid resistance to particles, calculations based on the apparent density must be performed.

1.2.6 Non Spherical Particles

Most atmospheric particles are non spherical. Therefore, it is necessary to convert these irregular particles into equivalent spherical particle size to correctly account for and use d and d_a relationships. This can be done quite simply using the dynamic shape factor (χ) and the equivalent diameter (d_e). Equivalent diameter is calculated using the actual particle volume (V) and

$$d_e = (6V/\pi)^{1/3}$$

The value of χ can be obtained using Hesketh's curve [2] as given in Figure 1.3. Finally, the Stoke's diameter is

$$d = d_e/\sqrt{\chi}$$

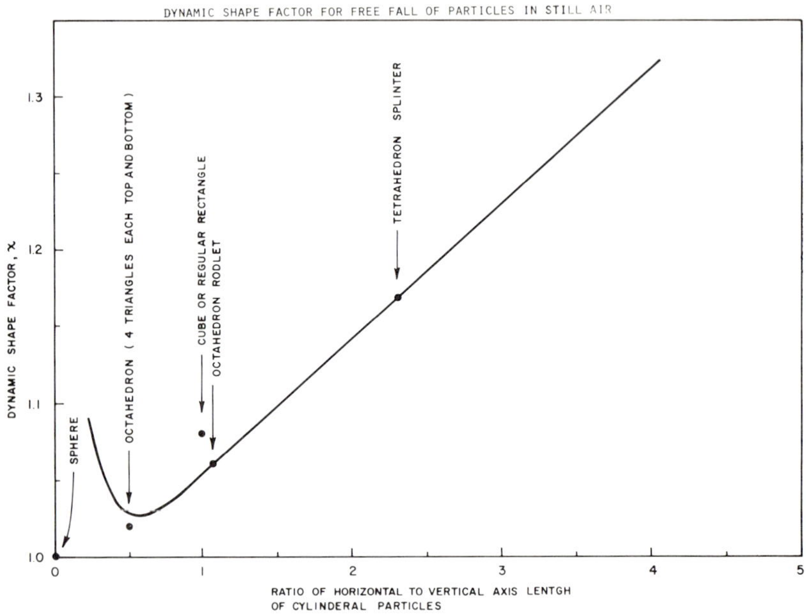

Figure 1.3. Dynamic shape factor for free fall of particles in still air.

1.2.7 Pressure Effects

The influence of pressure is usually not significant, however, there are times it can affect the data and should always be considered. Dr. V. A. Marple [6] developed the convenient relationship shown in Figure 1.4. Note that this can be used to relate all of particle diameter, pressure, P, and Cunningham connection factor, C. Tacitly implied is the relation between Stokes diameter, d, and aerodynamic diameter, d_a, by the inclusion of particle density, ϱ_p, as given in Section 1.2.3. The defining equation used in that section is rewritten for convenience as

$$d\sqrt{C}\sqrt{\varrho_p} = d_a\sqrt{C_a}$$

remembering that by definition aerodynamic particle density is unity. If each side of this equation is multiplied by pressure, P, the equality remains. The Marple curve applies consistently to *either:* aerodynamic $\sqrt{C}\ d_pP$ vs. d_pP; or to Stokes $\sqrt{C}\ d_pP$ vs. d_pP where d_p is general "particle diameter." It does not apply to combinations, yet it can be used to convert specifically from d_a to d. The following example shows how this is achieved.

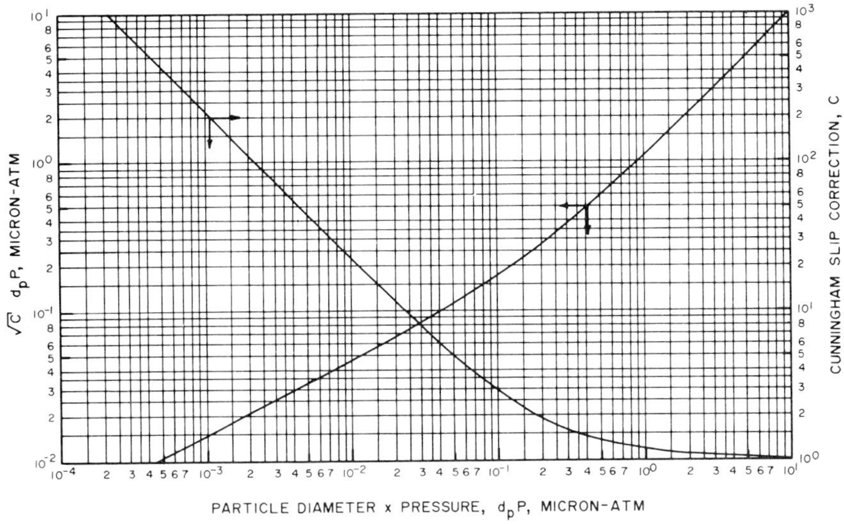

Figure 1.4. Marple curve for pressure—size—slip relationships (Courtesy Sierra Instruments, Inc. [6]).

Example 1.2

A particle with a density of 2 is measured by a size selective impactor in Denver as having an aerodynamic diameter of 0.20 μm. The pressure is 0.92 atmosphere. Determine Stokes diameter, C_a and C.

Solution:

$$\text{aerodynamic } d_pP = d_{pa}P = (0.20)\,(0.92) = 0.18 \text{ micron-atm}$$

From the correct curve in Figure 1.4 and the right side ordinate, find $C_a \cong 2.0$ (use caution reading these log coordinates).

Using the correct curve and the left side ordinate, find

$$\sqrt{C_a} d_{pa} P = 0.25$$

(Note that C_a can also be found by $\left(\frac{.25}{.18}\right)^2 = 1.9$.) Therefore

$$\sqrt{C_a} d_p P \sqrt{\varrho_p} = 0.25$$

and

$$\sqrt{C_a} d_p P = 0.25/\sqrt{2} = 0.18$$

From the figure, find the abscissa to be

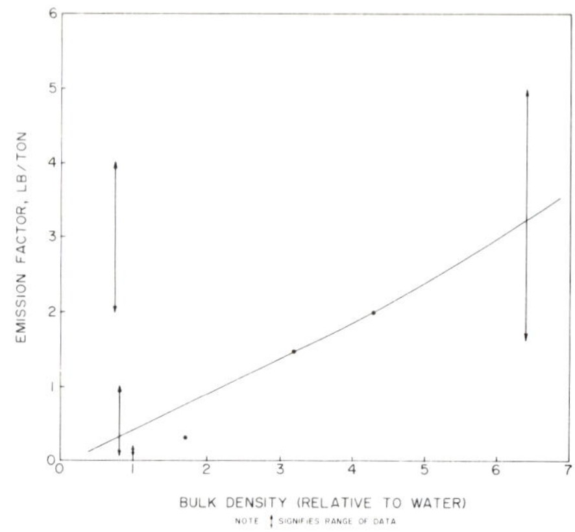

Figure 1.5. Misleading apparent relationship for emissions from transferring and conveying operations.

$$d_pP = 0.12 \text{ micron-atm}$$

and

$$d_p = 0.12/0.92 = 0.13\ \mu m$$

Finally, $C \cong 2.5$ (also equals $\left(\frac{.18}{.12}\right)^2 = 2.3$).

1.2.8 Transport vs. Type and Source

Fugitive dust losses to the atmosphere depend on numerous factors including material type (physical and chemical properties), and source effects such as wind speed, storage shape and/or arrangement, type of operation, and especially moisture current not only of the dust but the humidity of the air. Historical developments including process modifications and the incorporation of generic dust control on dusty operations can lead to interesting but misleading conclusions. For example, if one plots data such as that available for transfer and conveying operations [1] to show "uncontrolled" emission factor values as a function of bulk density it would appear that emissions are less for lower density materials. This misleading approximation is shown in Figure 1.5. Actually, unspecified factors as well as properties mentioned above all effect emissions and would yield the true reasons for this relationship. This section discusses some of these source related factors exclusive of the material type properties.

Wind—Wind speed, U, is one of the most important factors related to amount of fugitive emissions even though this effect can be rendered "insignificant" by proper protection of the pile (e.g. in the extreme, completely enclose the material). In addition to speed, frequency of wind speed over 12 mph, f, directly affects emissions.

Moisture—Water content is also a major factor in how much dust is lost and can be accounted for in various relationships as: average days of $\geqslant 0.01$ inch rainfall, various "precipitation-evaporation" indices, sample moisture content, M, and by relative humidity of the air. This is an indirect relationship in that the greater the amount of moisture the lower the emissions.

Silt—Fugitive emissions are directly related to the silt content, S, where silt is defined as amount of material passing through a 200 mesh screen.

Movement—Emissions are directly related to mechanical movement or disturbance of the dust and are functions of the applied force (e.g. vehicle weight, W), rate of movement, s, type of operation, O (e.g. loading, unloading, crushing, screening, road traffic, etc.) and amount moved. Emission factors based per mass of material are lower as more material is mechanically moved per single base of volume, Y.

Storage Duration—In addition to the silt variation, an old pile could have more adhesive force between particles to cement and hold them together or it could have a breakdown of material and greatly increased emissions. Thus storage duration could either increase, D_i, or reduce, D_r, emissions depending upon the physical and chemical nature of the material.

Storage Arrangement—If the material rises above the plane elevation of the surrounding landscape, the frontal slope angle, θ, greater than 10 degrees, the orientation of the material with respect to the wind direction, ϕ, and the material surface area, A, affect emissions.

Controls—These can consist of adjusting any one or more (or even all) of the factors discussed above by the factor K.

Summary—Using the descriptions just provided, a simplistic dust emission factor, EF, relationship per unit mass of material, is

$$EF \; \alpha \; \frac{KSs\,WOf\,D_i\,A^{1/3}\,\theta \sin \phi\, U^{1\ to\ 3}}{M^2\,Y\,D_r}$$

It is obvious that this certainly leaves many opportunities for reducing fugitive dust emissions.

1.2.9 Prediction of Dust Deposition

Sections 1.4.2 through 1.4.8 explain the different techniques that are currently employed for measuring the rate of dust deposition. Although these techniques are used for direct measurement of dust deposition, they are unable to provide size distribution because deposits contain all particle sizes. For estimates of deposition rate of particles of individual size range theoretical deposition models may be applied. However, these models require necessary information as an input data. This information includes particulate concentration at a known height along with size distribution, wind speed, atmospheric stability, and the equivalent roughness height of the deposition surfaces, too. Various models are available for estimating the deposition of particles from the atmosphere [7].

The parameter that measures the rate of particles deposit is "the deposition velocity, Vd" and is defined as follows:

$$Vd = \frac{\text{amount of particles deposited/unit time/unit area}}{\text{particle concentration}}$$

Once the deposition velocity for a given particle size is evaluated, the deposition flux is simply determined by multiplying the deposition velocity by the particle concentration as measured or predicted beforehand. Although the deposition flux may be approximately constant near the ground, the concentration will still be a function of height, as will the deposition velocity. Accordingly, it is the usual practice to define the deposition velocity at some small reference height.

Two deposition models are selected and will be described here. These models account for the surface roughness scale length, a parameter of great importance when investigating deposition of particulate over open areas of different surface characteristics (e.g., open, forested or urban area).

Atmospheric stability conditions are also considered in these models.

1.2.9.1 Estimation of Particulate Deposition Velocity

Model 1. Sheih et al. [8] have presented an approach for calculating the deposition of sulfate particles from the atmosphere. This approach can be adapted to be applied for the wide range of atmospheric particles. The deposition velocity is given by:

$$V_d = KU_* [\ln (Z/Zo) - \psi_c] + V_s$$

where

$K = 0.4$ (is the Von Karman constant)
U_* is the friction velocity (given later)
Z is the height at which the velocity and concentration are measured

Zo is the surface roughness scale length (defined later)
ψ_c is the stability correction function
V_s is the particle settling velocity (Section 1.2.2)

The stability correction function is calculated as follows:

$$\psi_c = \exp\{0.598 + 0.390 \ln(-Z/L) - 0.09 [\ln(-Z/L)^2]\}$$

where L is the Obukhov length, a function of the atmospheric stability class as given by Table 1.3.

Table 1.3 should not be used as a means of relating wind speeds to stability classes, but only as an explanation of the Obukhov relation.

Model 2. This model is proposed by El-Shobokshy and Sharaf Eldin [9], and it also accounts for the roughness scale of the deposition surface as well as the atmospheric stability. The particles deposition velocity can be estimated by using the following equations:

For unstable conditions:

$$V_d^* = \frac{V_s^*}{1-\exp\left\{\frac{V_s^*}{KX}\ln\left(\frac{D^* + KX\,Zo^*}{D^* + KXZ^*}\right)\right\}}$$

where

$X = (1 - 15\ Z/L)^{-0.25}$
$K = 0.4$
D = Molecular diffusivity of the particles (significant for submicron particles and neglected for particles larger than 1 μm diameter).

For approximation, value of L can be taken from Table 1.

For neutral conditions:

Table 1.3. Approximate Relationship between Stability Classes, Obukhov Length (L), and Wind Speeds for Flow Over Land.

Stability Class		Obukhov Length L,m	Windspeed m/s
Very unstable	(A)	- 8	1.5
Moderately unstable	(B)	-14	2.5
Slightly unstable	(C)	-40	4.0
Near Neutral	(D)	large	6.0
Moderately stable	(E)	40	3.5
Very stable	(F)	15	2.5

$$V_d^* = \frac{V_s^*}{1\text{-exp}\left\{\dfrac{V_s^*}{K}\ln\left(\dfrac{D^* + KX\,Zo^*}{D^* + KZ^*}\right)\right\}}$$

For slightly stable conditions:

$$V_d^* = \frac{V_s^*}{1\text{-exp}\dfrac{1 + 5.2L}{K}\ln\left\{\dfrac{D^* + (K/1 + 5.2/L)\,Zo^*}{D^* + (K + 1 + 5.2/L)\,Z^*}\right\}}$$

For stable conditions:

$$V_d^* = \frac{V_s^*}{1\text{-exp}\left\{\dfrac{5.2}{D}V_s\ln\left[\dfrac{D^* + (K/5.2)\,Zo^*}{D^* + (K/5.2)\,Z^*}\right]\right\}}$$

Table 1.4 gives different land-use types and the associated estimated surface roughness size, Zo, value that can be used for both deposition models described above.

In the above equations:

$$V_d^* = V_d/U_*$$
$$V_s^* = V_x/U_*$$

$$D^* = D/V \text{ (V = Kinematic Viscosity of air)}$$

Table 1.4. Land-Use Types and the Associated Values of Surface Roughness, Zo (cm).

Land-Use Type	Zo (cm)
Cropland and Pasture	20
Cropland, Woodland, and Grazing Land	30
Irrigated Crops	5
Grazed Forest and Woodland	90
Ungrazed Forest and Woodland	20–100
Subhumid Grassland and Semiarid Grazing Land	10
Open Woodland Grazed	20
Desert Shrubland	30
Swamp	20
Marshland	50
Metropolitan City	50–1000
Lake or Ocean	0.01

The rate of particulate deposition per unit area N can be estimated by:

$$N = V_d \cdot C_o$$

where C_o is the particulate concentration above the surface. The total flux, $\bar{N}$ from the atmosphere containing polydispersed particles of size range o—d_p and with a frequency distribution of $f(d_p)$ may then be calculated from

$$\bar{N} = \int_o^{d_p} N\,(d_p)\, f(d_p)\, d(d_p)$$

The friction velocity, U_*, depends on the local wind speed, surface roughness, and the atmospheric stability as given in the next section.

1.2.9.2 The Friction Velocity in the Surface Layer

Considering the variation of wind with height in the surface layer of the atmosphere (up to 30 to 60 m). The friction velocity of the surface layer, U_*, is calculated as follows:

For unstable conditions:

$$U_* = \bar{U}\, K / \left\{ 2\,(\tan^{-1}\xi - \tan^{-1}\xi_1) + \ln\left(\frac{\xi - 1}{\xi_1 - 1}\right) - \ln\left(\frac{\xi + 1}{\xi_1 + 1}\right)\right\}$$

where $\bar{U}$ = prevailing wind speed at a height Z

$$\xi = \left[1 - 15\left(\frac{Z + Zo}{L}\right)\right]^{0.25}$$

$$\xi_1 = \left[1 - 15\left(\frac{Zo}{L}\right)\right]^{0.25}$$

For neutral conditions:

$$\bar{U}_* = U\, K / \ln \frac{Z + Z_o}{Z_o}$$

For slightly stable conditions:

$$\bar{U}_* = U\, K / \quad \ln \frac{Z + Z_o}{Z_o} \quad + 5.2\, Z/L$$

For stable conditions:

$$U_* = U\ K/\{\ln \frac{Z+Z_o}{Z_o} + 5.2\}$$

1.2.9.3 Open Terrain Deposition Velocity

The model of El-Shobokshy and Sharaf El-Din has been used for estimating the deposition velocity of particulate material of density 1.6 g/cm^3 on surfaces with roughness heights of

$$Zo = 5,\ 0.80 \text{ and } 0.04 \text{ m}$$

These values represent urban, forested, and open areas. The atmospheric stabilities are considered in the model and Figure 1.6 shows the deposition velocity of dust particles at different wind speeds and surface roughness for unstable atmosphere.

In Figure 1.6, a plot for settling velocity as a function of particle diameter is shown. For neutral atmospheric stability, the results are similar to those shown in Figure 1.6. For stable atmospheres, particles in the size range of 1 - 100 μm fall out from the atmosphere at about their settling velocity with nearly negligible effect due to surface

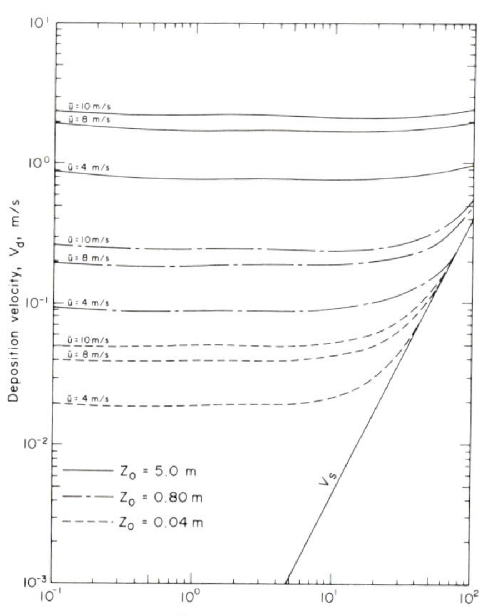

Figure 1.6. Deposition velocity of dust particles as a function of surface roughness for unstable atmosphere at different wind speeds.

roughness as shown in Figure 1.7. For neutral and unstable atmospheres, both surface roughness and wind speed does have appreciable effect on the dust fall velocity. In the model, it has been shown that the ambient temperature has a nearly nonexistent effect on the velocity of deposition.

1.3/Particulate Dispersion Models

Upon emission from the source, the fugitive dust disperses by atmospheric diffusion. Estimation of atmospheric particulate concentration in a remote area due to a specific source is of great concern. To assess the impact of a given source, various dispersion models have been established and applied for different types of sources. Slade [10] and Turner [11] cover the transport and diffusion of gases and particulates (less than 20 μm) using diffusion model developed with the following assumptions:

1. The plume spread has a Gaussian distribution in both the horizontal and vertical planes with standard deviations of plume concentration distribution in the horizontal and vertical of σ_y and σ_z respectively.
2. The mean wind speed affecting the plume is U.
3. There is no deposition or reaction at the earth's surface.

The concentration, C, of particulates at x, y, z (Figure 1.8), from a con-

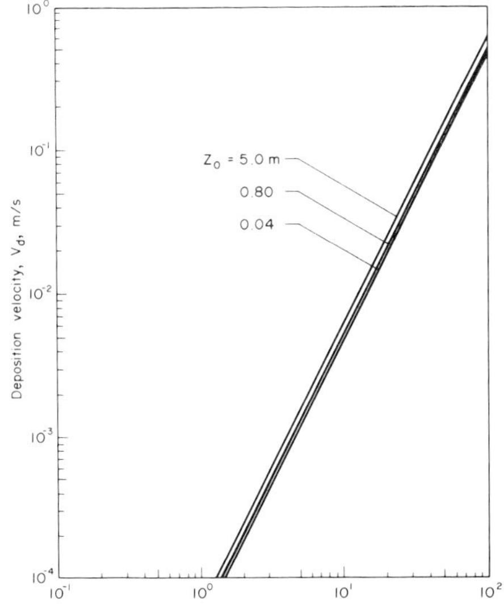

Figure 1.7. Deposition velocity of dust particles from stable atmosphere at 10 m/s wind speed.

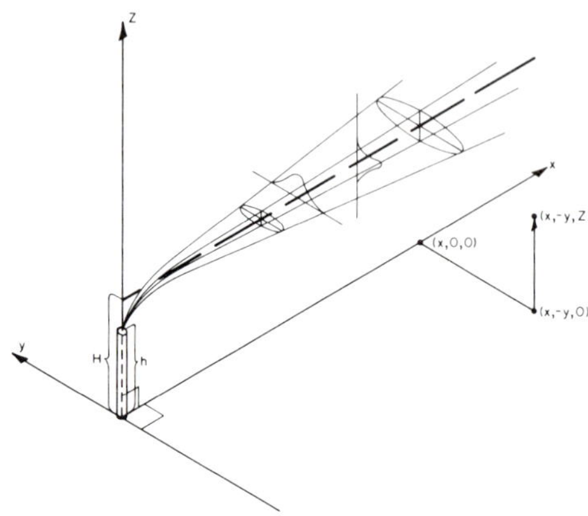

Figure 1.8. Coordinate system showing plume downwind distributions in the horizontal and vertical planes.

tinuous source of strength, Q, with an effective emission height, H, is given by:

$$C_{(x,\, y,\, z,\, H)} = \frac{Q}{2\,\pi\,\sigma_y\,\sigma_z\,U}\exp\left[-\tfrac{1}{2}\,(y/\sigma_y)^2\right]$$

$$\left\{\exp\left[-\tfrac{1}{2}\left(\frac{Z-H}{\sigma_z}\right)^2\right] + \exp\left[-\tfrac{1}{2}\left(\frac{Z+H}{\sigma_z}\right)^2\right]\right\}$$

The standard deviations are obtained from Figures 1.9 and 1.10 as a function of distance downwind and the atmospheric stability class as a parameter. Sources are classified in three main categories, namely, point, line, and area source. However in real situations, dispersion patterns must be established for complex sources that release several pollutants simultaneously. Each of these are discussed in the following sections.

1.3.1 Dispersion from Point, Line, and Area Sources

Dispersion of particulate material is not only dependent on the source strength, location, wind speed, and atmospheric stability, but also depends on the geometry of the source. The source may be approximated as point, line, or area source. Examples of different types of sources according to this classification along with the relevant dispersion equations are given in Table 1.5.

Application of the dispersion equation requires values of the plume standard deviations σ_y and σ_z. These values are dependent on the at-

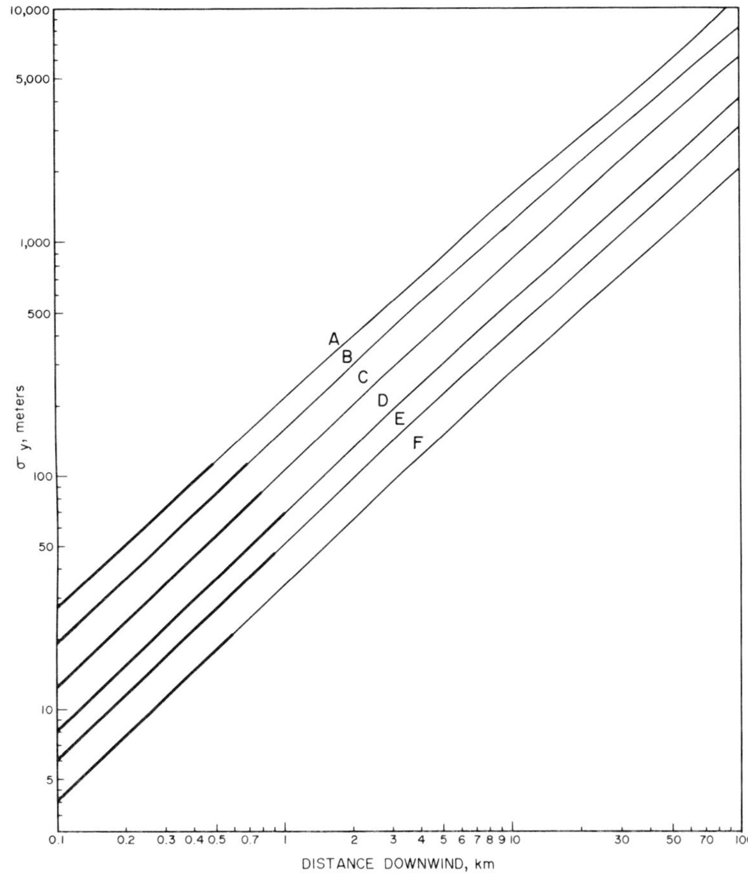

Figure 1.9. Horizontal dispersion coefficients as a function of downwind distance from the source [11].

mospheric stability class. The following section describes a procedure for determining the stability class.

1.3.2 Determination of Atmospheric Stability Class

Two methods may be used to determine the atmospheric stability. The first is the Pasquill [12] stability category and this assumes that sampling time over which the concentration is to be determined is 10 min, and the wind speed is measured at the height 10 m from the ground. The stability class depends on wind speed, cloud cover, solar radiation, and day or night as given in Table 1.6.

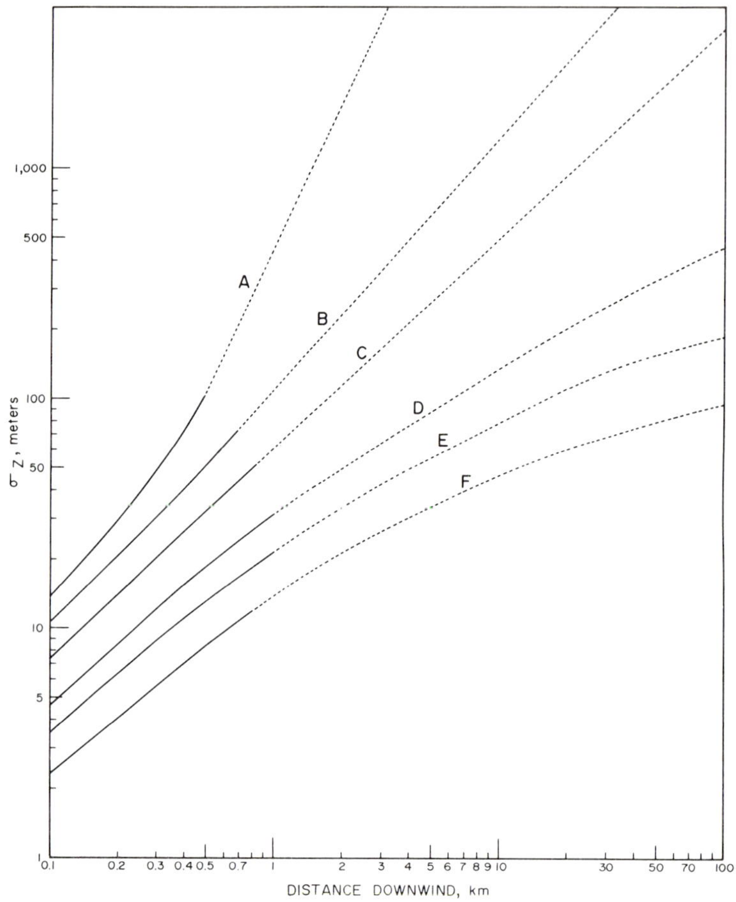

Figure 1.10. Vertical dispersion coefficients as a function of downwind distance from the source [11].

The stability class can also be determined by another method in which the standard deviation of the horizontal wind direction σ_θ is calculated. It has been recommended [13] that σ_θ represent a one-hour period and include important wind direction factors. Also, the root-mean square of the overall mean wind direction for the sampling period should be used. Following Pasquill's recommendations summarized in Table 1.7, Carney and Coblenz [14] observed the dependence of hourly standard deviations of the wind direction on wind meander and presented a method for determining σ_θ. Table 1.7 outlines this method. Table 1.8 gives the atmospheric stability class as a function of σ_θ ranges.

Table 1.5. Dispersion Equations as Applied to Different Type of Sources.

Type of Source	Example of Sources	Source Location	Dispersion Equation
Point	Stack, crushing, screening, etc.	stack height H	a) along plume C_L: $C = (Q/\pi\sigma_y \sigma_z U) \exp[-½ (H/\sigma_z)^2]$ b) off C_L by distance y: $C = (Q/\pi\sigma_y \sigma_z U) \exp[-½ (H/\sigma_z)^2] \exp[-½ (y/\sigma_y)^2]$
	loading, drilling, etc.	ground level	a) along plume C_L: $C = Q/\pi\sigma_y \sigma_z U$ b) off C_L by distance y: $C = (Q/\pi\sigma_y \sigma_z U) \exp[-½ (y/\sigma_y)^2]$
Line	material line pile	stack height H	a) wind normal to the line $C = (2Q/\sqrt{2}\, \sigma_z U) \exp[-½ (H/\sigma_z)^2]$ b) wind at an angle ϕ with the line $C = (2Q/\sin\phi \sqrt{2\pi\sigma_z}\, U) \exp[-½ (H/\sigma_z)^2]$
	haul roads, scarpers . . . etc.	ground level	a) wind normal to the line $C = 2Q/\sqrt{2\pi}\, \sigma_z U$ b) wind at an angle d with the line $C = 2Q/\sin d \sqrt{2\pi}\, \sigma_z U$
Area	agricultural land, material area piles, large number of print sources in close vicinity.	at a height H or ground level	Proceed as follows: 1. Approximate the area source as a square with a side S 2. Assume an initial standard deviation $\sigma_{y_0} \cong S/4.3$ 3. Use Figure 1.10 to determine the distance X_z that gives the value of σ_{y_0} 4. Calculate the effective distance from the source as $X_{eff} = X_{actual} + X_a$ 5. Determine σ_y (Figure 1.9) as a function of X_{eff} 6. Apply the equation of point source

1.3.3 Complex Source Models

Table 1.5 gives the fundamental dispersion equations as applied to point, line, and area sources. However, in most air pollution investigations complex sources and configurations including the effect of building downwash. Turner [15] presented a critical review as of 1979 which lists many of these atmospheric dispersion models. Table 1.9 up-

Table 1.6. **Key to Stability Categories [11].**

Surface Wind Speed at 10 m height (m/sec)	Insolation Stability Classes				
	Day			Night	
	Strong[a]	Moderate[b]	Slight[c]	Thinly Overcast or > ½ Cloud[d]	Clear to < ½ Cloud
<2<(4.5 mph)	A	A-B	B	—	—
2-3 (4.5-6.7)	A-B	B	C	E	F
3-5 (6.7-11)	B	B-C	C	D	E
5-6 (11-13.5)	C	C-D	D	D	D
>6 (>13.5 mph)	C	D	D	D	D

[a]Sun >60° above horizontal; sunny summer afternoon; very convective.
[b]Summer day with few broken clouds.
[c]Sunny fall afternoon; summer day with broken low clouds; or summer day with sun from 15-35 C with clear sky.
[d]Winter day.
Insolation = amount of sunshine.

A: Very unstable
B: Unstable
C: Slightly Unstable
D: Neutral
E: Slightly Stable
F: Stable

Table 1.7. Determination of Hourly $(\sigma_\theta)_{hr}$.

Parameter	Method of Determination
σ_L	1. Sample wind direction* every 2.5 min. 2. Sample time one hour (24 sample) 3. Calculate mean value over one-hour = σ_L
σ_H	1. Sample wind direction every 5 seconds 2. Use 65 sample and calculate 5-min Stan. Dev. σ_5 3. Calculate the mean value of 12 $\sigma_5 = \sigma_H$
$(\sigma_\theta)_{hr}$	$(\sigma_\theta)_{hr} = (\sigma_L{}^2 + \sigma_H{}^2)^{1/2}$

*A low threshold vane type wind system is recommended. The atmospheric stability is then determined on hourly basis according to Table 1.8.

Table 1.8. Atmospheric Stability Class as a Function of σ_θ.

σ_θ	Stability Class
$\sigma_\theta > 22.5°$	A
$17.5 < \sigma_\theta < 22.5°$	B
$12.5 < \sigma_\theta < 17.5°$	C
$\sigma_\theta < 12.5°$	D

Table 1.9. Available Air Pollution Models.

Model	Applicability	User's Guide
APRAC	Stanford Research Institute's urban carbon monoxide model. Computes hourly averages for any urban location. Requires an extensive traffic inventory for the city of interest.	User's Manual for the APRAC-1A Urban Diffusion Model Computer Program (NTIS accession number PB 213-091)
CDM	The Climatological Dispersion Model determines long term (seasonal or annual) quasi-stable pollutant concentrations at any ground level receptor using average emission rates from point and area sources and a joint frequency distribution of wind direction, wind speed, and stability for the same period.	A. D. Busse and J. R. Zimmerman, "User's Guide for the Climatological Dispersion Model," U.S. Environmental Protection Agency, Research Triangle Park, NC, Environmental Monitoring Series, EPA-R4-73-024, 1973 (NTIS accession number PB 227-346).
HIWAY	Computes the hourly concentrations of nonreactive pollutants downwind of roadways. It is applicable for uniform wind conditions and level terrain. Although best suited for at-grade highways, it can also be applied to depressed highways (cut sections).	J. R. Zimmerman and R. S. Thompson, "User's Guide for HIWAY: A Highway Air Pollution Model," U.S. Environmental Monitoring Series, EPA-650/4-74-008, 1975 (NTIS accession number PB 239-944)
PTMAX	Performs an analysis of the maximum short-term concentrations from a single point source as a function of stability and wind speed. The final plume height is used for each computation.	
PTDIS	Estimates short-term concentrations directly downwind of a point source at distances specified by the user. The effect of limiting vertical dispersion by a mixing height can be included and gradual plume rise to the point of final rise is also considered. An option allows the calculation of isopleth half-widths for specific concentrations at each down-wind distance.	D. B. Turner and A. D. Busse, "User's Guide to the Interactive Versions of Three Point Source Dispersion Programs: PTMAX, PTDIS, and PTMTP." Preliminary Draft, Meteorology Laboratory, U.S. Environmental Protection Agency, Research Triangle Park, NC, 1973.
PTMTP	Estimates for a number of arbitrarily located receptor points at or above ground-level, the concentration from a number of point sources. Plume rise is determined for each source. Downwind and crosswind distances are determined for each source receptor pair. Concentrations at a receptor	

(continued)

Table 1.9. (continued).

Model	Applicability	User's Guide
	from various sources are assumed additive. Hourly meteorological data are used; both hourly concentrations and averages over any averaging time from one to 24 hours can be obtained.	
CDMQC	The algorithm is the Climatological Dispersion Model (CDM) altered to provide implementation: of calibration, of individual point and area source contribution lists, and of averaging time transformations. The basic algorithms to calculate pollutant concentrations used in the CDM have not been modified, and results obtained using CDM may be reproduced using the CDMQC.	Kenneth L. Brubaker, Polly Brown, and Richard R. Cirillo Addendum to User's Guide for Climatological Dispersion Model. Prepared by Argonne National Laboratory for the U.S. Environmental Protection Agency, Research Triangle Park, NC, EPA-450/3-77-015, 1977 (NTIS accession number PB 274-040).
CRSTER	This algorithm estimates ground-level concentrations resulting from up to 19 colocated elevated stack emissions for an entire year and prints out the highest and second highest 1-hr, 3-hr, and 24-hr concentrations as well as the annual mean concentrations at a set of 180 receptors (5 distances by 36 azimuths). The algorithm is based on a modified form of the steady-state Gaussian plume equation which uses empirical dispersion coefficients and includes adjustments for plume rise and limited mixing. Terrain adjustments are made as long as the surrounding terrain is physically lower than the lowest stack height input. Pollutant concentrations for each averaging time are computed for discrete, non-overlapping time periods (no running averages are computed) using measured hourly values of wind speed and direction, and estimated hourly values of atmospheric stability and mixing height.	Monitoring and Data Analysis Division, User's Manual for Single Source (CRSTER) Model. U.S. Environmental Protection Agency. Research Triangle Park, NC. EPA-450/2-77-013, 1977 (NTIS accession number PB 271-360)
PAL	Point, Area, Line source algorithm. This short-term Gaussian steady-state algorithm estimates concentrations of stable pollutants from point, area, and line sources. Computations from area	William B. Petersen, User's Guide for PAL—A Gaussian-Plume Algorithm for Point, Area, and Line Sources. U.S. Environmental Protection Agency, Research Triangle Park, NC,

(continued)

Table 1.9. (continued).

Model	Applicability	User's Guide
	sources include effects of the edge of the source. Line source computations can include effects from a variable emission rate along the source. The algorithm is not intended for application to entire urban areas but for smaller scale analysis of such sources as shopping centers, airports, and single plants. Hourly concentrations are estimated and average concentrations from 1 hour to 24 hours can be obtained.	Environmental Monitoring Series EPA-600/4 78-013, 1978 (NTIS accession number PB 281-306).
VALLEY	This algorithm is a steady-state, univariate Gaussian plume dispersion algorithm designed for estimating either 24-hr or annual concentrations resulting from emissions from up to 50 (total) point and area sources. Calculations of ground-level pollutant concentrations are made for each frequency designed in an array defined by six stabilities, 16 wind directions, and six wind speeds for 112 program-designed receptor sites on a radical grid of variable scale. Empirical dispersion coefficients are used and includes adjustments for plume rise and limited mixing. Plume height is adjusted according to terrain elevations and stability classes.	Edward W. Burt, "Valley Model User's Guide," U.S. Environmental Protection Agency, Research Triangle Park, NC. EPA 450/2-77-018, 1977 (NTIS accession number PB 274-054)
RAM	Gaussian-Plume Multiple Source Air Quality Algorithm. This short-term Gaussian steady-state algorithm estimates concentrations of stable pollutants from urban point and area sources. Hourly concentrations and averages over a number of hours can be estimated. Briggs plume rise is used. Pasquill-Gifford dispersion parameters thought to be valid for urban areas are used. Concentrations from area sources are determined using the method of Hanna, that is, sources directly upwind are considered representative of area source emissions affecting the receptor.	D. Bruce Turner and Joan Hrenko Novak, "User's Guide for RAM," U.S. Environmental Protection Agency, Research Triangle Park, NC. EPA-600/8-78-016, November 1978.

(continued)

Table 1.9. (continued).

Model	Applicability	User's Guide
	Special features include determination of receptor locations downwind of significant sources and determination of locations of uniformly spaced receptors to ensure good area coverage with a minimum of receptors.	
ISC	Evaluate impact of emission from wide variety of sources found at large industrial source complexes. Site specific meteorological data can be input for research purposes. Range of applicability: from a single isolated stack to groups of source complexes containing both traditional and non-traditional pollutant sources.	Industrial Source Complex (ISC) Dispersion Model User's Guide (Vol. I), EPA-450/4-79-30, Environmental Protection Agency, Research Triangle Park, NC (1979), J.F. Bowers, et al.
BLP	For Modelling multiple Buoyant Line and Print Sources. For each point source, the stock location, base elevation, physical stack height, inside stack diameter, stack gas exit velocity, stack gas temperature and pollutant emission rate must be included as input. Up to 50 point source are allowed. For each line source the coordinates of the line end points, base elevation, buoyancy flux, building height, roof vent width, building separation, and pollutant emission rate are required. A maximum of 10 line sources is allowed.	Shulman, L. L. and J. S. Scine 1980. Buoyant Line and Point Source (BLP) Dispersion Model User's Guide. Document P-7304B. Environmental Research of Technology, Inc. Concord, MA

dates and expands Turner's listing and presents many available dispersion models for complex sources along with their applicability and a user's guide to these models.

1.4/Deposition of Dust

1.4.1 Introduction

The deposition of dust particles is considered as the main mechanism by which airborne pollution is removed from the atmosphere. Dust particles emitted from a given source are dispersed in the atmosphere where the small particles stay in suspension for a long time while the larger particles settle towards the earth under gravity at near distances from the source. The rate of deposition of dust on smooth surfaces is given by the concentration above the surface times of the settling velocity of particles. This gives the particle flux in m mg^{-2} sec^{-1}. For a given rate of deposition the concentration will be proportional to the wind speed. As a result, the rate of deposition of small particles is inversely proportional to wind speed and directly proportional to particle settling velocities. On the other hand, the settling velocity of large particles dominates movement by diffusion, and the maximum deposition takes place near a source. Further, the location of maximum deposition of large particles moves nearer towards the source as the wind speed decreases. It may, therefore, be noticed that the maximum deposition rates increase more rapidly as the wind speed falls and increase of particle settling velocity than for smaller particles.

As far as the size is concerned, the large and small particles have not been well defined in the foregoing discussion as there is not a specified boundary between the two size ranges. However, particles larger than 20 μm are usually deposited at their settling velocities within short distances from the source in most rock industries for a wide range of wind speeds. Small particles, therefore, may be considered as those smaller than 15–20 μm.

The deposition of dust is known to be one of the adverse effects that is associated with dust emission. The result of deposition which is referred to as a "deposit" has been usually measured by a standard "deposit gauge" or an equivalent device of similar principle. Investigations of fugitive dust necessitate the measuring deposition rates in order to produce a clear evidence about the impact of particulate pollution. The next section describes the method of deposition measurement along with the necessary devices.

1.4.2 Principles of Deposition Measurements

As a standard technique, the deposition is usually measured by using an open-mouth container having a circular orifice known by a dust deposit gauge that is exposed to the atmosphere for a sufficient period of time (usually one month) to collect a representative amount of deposit. Most of the reported data indicate that dust fall measurement is not precise at all times. However, the deposit gauge gives only a reasonable measure of deposition on the earth's surface when wind speeds are low. In strong winds the deposit gauges catch very little dust, and even the amount caught is appreciably affected by any change in the vertical angle between the wind direction and the plane of the gauge. Further reduction in the catch is expected because the wind lifts as it passes over the gauge.

In the following sections different methods available for measuring the dust fall are presented along with associated instruments and the measuring procedure. The choice of the measuring method in a specific investigation depends on the local conditions, the length of the possible measuring period, dust loading, the required analysis of the collected dust, e.g. chemical, morphological, and the special requirements of the studies being made.

1.4.3 The Dust Fall Gauge

There are some different designs of the dust fall gauge having the same principle. The most traditional one is the British Standard Gauge [16] shown in Figure 1.11. It consists of a trapping funnel made of glass with an upper diameter 31 cm, trapping area 760 cm². The funnel outlet is sealed and fitted to a 10 liter bottle "collector" having dimensions of 18.5 cm diameter × 21 cm high. A wide-mesh is mounted on top of the device as a protection against birds, leaves, etc. The dust fall gauge is placed in a 150 cm high tripod.

The bottle "dust collector" must be filled to about 75 percent of its capacity with distilled water. It is recommended to add 0.10 g of reagent grade mercuric chloride. This will act as both an algacide and fungicide. If there is a possibility that the collecting liquid may be consumed by

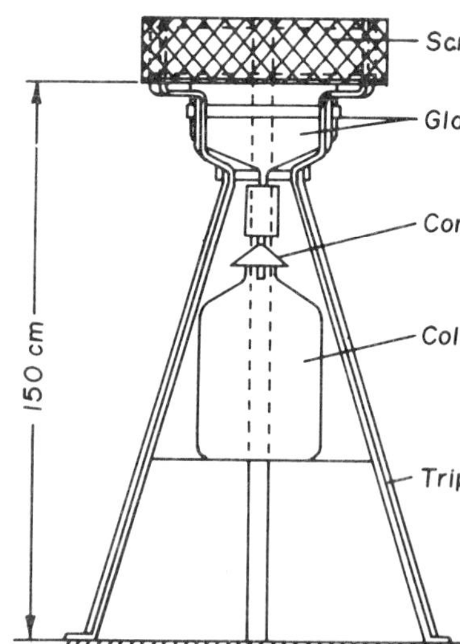

Figure 1.11. British standard dust fall gauge.

humans or animals, 0.10 g of copper sulfate pentahydrate shall be used as an algae inhibitor. This has no fungicidal properties. If in dry weather area the rate of evaporation is too high, make up water should be added to keep the liquid in its normal level.

1.4.4 Sampling Procedure

1. First, it is important to choose the proper sampling site. If the investigation is concerned with some specific source and it is required to quantify the dust deposits over a given area, therefore, dust fall gauge must be arranged in the upwind and downwind of the source if the prevailing wind has nearly a defined direction. This is to enable the background dust to be determined and hence the actual amount of deposit due to the source only. However, if the prevailing wind changes the direction frequently, meteorological data should be recorded during the test to make the necessary adjustment during the analysis of results. The number of the dust fall gauge must be sufficient in a way to conform with the investigation objectives and to make allowance for duplicate readings whenever necessary.

The top of the dust fall gauge shall be a minimum of about 2 m and a maximum of 15 m above the ground or 1.5 m above any other surface such as a roof. Higher objects nearby the gauge must not be more than 30° from the horizontal i.e. a line drawn from the sampling container to the nearest edge of the highest point on any building should form not more than 30° angle with the horizontal.

It has also to be noted that the influence of a source close to the dust fall gauge can dominate the results. Therefore, after selecting the site, detailed notes should be made on the direction and distances of all near sources such as stacks, parking lots, unpaved roads, material storage piles . . . etc. Care must be made to insure the stability of the tripod and the whole gauge during any possible strong wind.

2. Having filled the bottle with the proper amount of liquid, a cover should be placed on the top and only be exposed when the dust fall gauge is placed in its holder at the sampling location. The adequate exposure period for obtaining a representative sample is 30 days ±3. However, some other techniques of deposit measurement are available that need only 1–2 weeks (Sections 1.4.7 and 1.4.8).

1.4.5 Analysis of Dust Fall Measurement

The collecting bottle has to be covered, sealed, and then transferred to the laboratory. The bottle must be scrubbed or stirred rigorously to remove all adhering materials. The contents, then poured in a clean container while screened using a 20-mesh screen, to remove any large stuff that may fall in during the testing period. Next, the content has to be filtered using filter paper or membrane filter to separate the precipitated dust (insoluble materials). The liquid is then evaporated to dryness in a weighed dish and then reweighed to determine the weight of soluble materials. The insoluble dust is dryed and the weight of which is determined. The total dust fall may then be determined by adding the soluble and insoluble amounts. The dust fall rate (deposition flax) is usually determined in terms of grams per square meter per month (based on 30 days). Thus

$$\text{Dust fall} = \frac{\text{Total Weight of dust}}{\text{Open area of the gauge}} \times \frac{30}{\text{time of exposure days}}$$

The dust fall may also be expressed as tons/sq. mile/month.

1.4.6 Polyethylene Dust Fall Collector

This is a simpler dust fall gauge compared to the standard gauge described in section 1.4.3. It consists of (Figure 1.12) a perforated polished aluminum open box used to support a polyethylene cup which is employed for the actual measurement of collected solids. The aluminum box has two functions: it reflects solar heat to evaporate water that may collect in the sample cup (rain, precipitation, etc.) and it acts as a wind break, thus minimizing the loss of samples due wind action. The perforations provide for the drainage of rain water and they minimize wind eddys. The polyethylene vessels are sufficiently light that they can be weighed directly on an analytical balance. Furthermore, they are not damaged by freezing conditions nor are they subject to corrosion and consequent changes in weight. The tared plastic container is placed in the aluminum box and left exposed for an appropriate period of time (usually a month) after which the container is covered with a matching lid and returned to the laboratory for subsequent analysis and tests. The moisture content can be evaporated on a water bath or by exposure to infrared radiation. Nonsignificant debris such as bugs, leaves, or twigs should be removed and the amount of residue then determined by heating the container and its contents at 110 °C for 1 hour, reweighing, and getting the dust fall by difference. The results may then be expressed as a weight of dust per unit timer per unit surface area. A liquid may be used in this collector to prevent the reentrainment of the collected dust by wind action, also the collector may have a greasy coating.

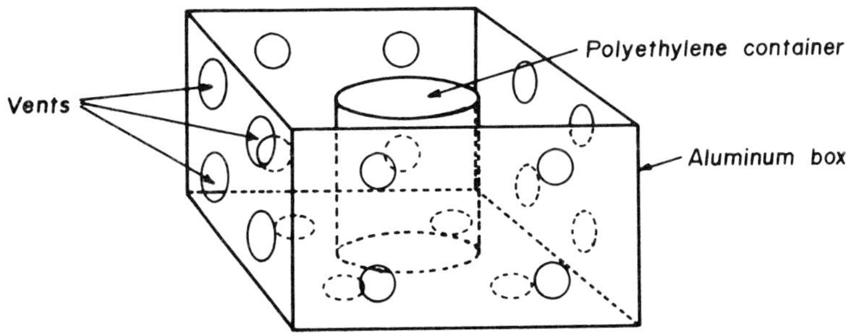

Figure 1.12. Polyethylene dust fall collector.

1.4.7 Diem's Adhesive Foil Instrument

This instrument [17] consists of a 0.07 mm aluminum foil having a surface area of 4.0 × 8.25 cm (33 cm^2 trapping area) and a frame. Pure vaseline (about 50 mg) or any equivalent grease is uniformly applied over the foil. The foil is warmed until the adhesive layer melts and then dried at 40 °C, cooled in a desiccator at a certain water vapor pressure and weighed. After exposure time the foils are collected, dried in the desiccator, and weighed. The maximum deposit onto the foil surface should not exceed 30 mg. The exposure time must then be estimated by the skill and the gained experience, which is generally fairly short. This method has a privilege of the possibility of inspection of the deposited dust under a reflection microscope for morphological tests.

The main disadvantage of Deim's method is that during heavy rain and intense insulation appreciable losses occur, so that the values obtained are frequently much lower than those obtained by methods employing dust fall gauges. The method has another disadvantage, namely, that it must be frequently repeated for long-term experiments. This method is, therefore, not recommended for determining a reliable and comparable dust level, but rather for testing intermittent local dust conditions at several measuring stations in order to detect dust emittents.

1.4.8 Plastic Petri Dish Method

For a short measurement duration of dust fall, and also when a microscopical examination for the collected dust is needed, the plastic Petri dish is available. In this method a circular transparent plastic Petri dish is used as a collecting device. The dish may have a desirable size (normally is about 20 cm diameter) having a side edge of about 2 cm high. A thin layer of vaseline or an equivalent kind of grease is applied on its surface and then placed in a drying device such as an oven with a temperature of about 40 °C for at least 4–5 hours, until it becomes completely dry and the greasy coating evenly spread on its surface. The Petri dish must then be covered and sealed. The cover may have a foam rubber edge for sealing and be secured to the dish by means of a strap or string.

In the measuring site the Petri dish may be placed on a stand about 1.20–1.50 m from the earth's surface. The principles of choice of the sampling site are the same as those described in Section 1.4.2. To start the measurement, the Petri dish is uncovered to the required length of time after which the cover is placed on the dish and transferred carefully to the laboratory. It is recommended to place in every sampling location from 2 to 3 dishes to check the duplication of results and to take the average for improving the accuracy.

The dishes must then be dried in a similar way as before the test and by means of an analytical balance, the weight of the collected dust in each Petri dish is determined. The transparent Petri dish has the advantage in that it can be used as a dust particles carrier under an optical microscope for morphological test and having photographs of the sample.

References

1. Hesketh, H. E. and Cross, F. L., Jr. *Fugitive Emission and Controls.* Ann Arbor Science Publishers (1982).
2. Hesketh, H. E. *Air Pollution Control.* AAS, 2nd printing (1981).
3. *TST Quarterly, V* (2), TSI Incorporated (1979).
4. Hesketh, H. E. *Fine Particulate in Gaseous Media.* Ann Arbor Science Publishers, 2nd Printing (1979).
5. Hesketh, H. E. *Air Pollution Control.* Ann Arbor Science Publishers, 2nd Printing (1981).
6. Marple, V. A. *Impactor Calculator.* Sierra Instruments, Inc., Carmel Valley, CA (1978).
7. Sehmel, G. A., "Particle Dry Deposition Velocities," Proceedings Atmosphere-Surface Exchange of Particulate and Gaseous Pollutants, Symposium Edited by Englemann, R. I. and Sehmel, G. A., CONF-740921, AEC Symposium Series, Oak Ridge, TN (1974).
8. Sheih, C. J., Wesely, M. L., and Hicks, B. B., "Estimated Dry Deposition Velocities of Sulfur Over the Eastern United States and Surrounding Regions," *Atmos. Environ. 13,* 1261–1368 (1979).
9. El-Shobokshy, M. S. and Sharaf El-Din, A., "Estimation of Dust Deposits from An Elevated Source Over Open Area at Different Atmospheric Stabilties," 8th International Clean Air Conference, Melborne (May 6–11, 1984).
10. Slade, D. H. *Meterology and Atomic Energy.* U.S. Atomic Energy Commission (1968).

11. Turner, D. B. *Workbook of Atmospheric Dispersion Estimates.* Publ. No. AP-26. U.S. EPA (1970).
12. Pasquill, F., "The Estimation of the Dispersion of Windborne Material," *Meterol. Mag., 90,* 1063, 33–49.
13. Hanna, S. R., et al., "AMS Workshop on Stability Classification Schemes and Sigma Curves—Summary of Recommendations," Bulletin AMS 1305–1309 (1977).
14. Carney, M. V. and Coblenz, J. G., "Experimental Observations of the Dependence Hourly Standard Deviation of the Wind Direction on Wind Meander," 5th Symposium on Turbulence, Diffusion, and Air Pollution, Atlanta (March 11, 1981).
15. Turner, D. B., "Atmospheric Dispersion Modeling. A Critical Review," *JAPCA, 29* (5), 502–519 (May 1979).
16. British Standard No. 1747, Part 1 "The Standard Deposit Gauge" (1969).
17. Deim, M., *Staub, 21,* 345 (1961).

2/Sampling and Monitoring

2.1/Introduction

One of the main objectives of fugitive dust investigations is to quantify the emission of particulates and relate the emission rates to the process rates. Emphasis is usually focused on emission factors of dust in the inhalable size range, i.e., less than 15 μm, aerodynamic diameter. Further, in order to meet the Clean Act Amendment requirement, it is required to refine the IP (inhalable particulate) data base into respirable "fine" <2.5 μm—and "coarse" 2.5 to 15 μm fractions. It is therefore important to select the proper equipment that provide the necessary data on which to base the dust emission results. Such equipment includes samplers for determining the mass concentration and the size distribution of suspended particulates.

Most fugitive dust investigations usually include the calculations of particulate dispersions as one of their main objectives or as a part of emission factor calculations. The dispersion models require meteorological data relevant to each particulate sampling process. It may be required to use meteorological instrumentations along with the particulate sampling equipment in order to obtain a consistent set of results.

The selector of equipment must consider the degree to which it enables the test method to meet the following requirements: (a) is it precise and accurate; (b) is it sensitive and specific; (c) is it economical; and (d) is it accepted, e.g. by the EPA? The following sections present the descriptions and major features of equipment currently used in the particulate emission studies. As mentioned previously, this equipment falls into two main categories: particulate sampling and meteorological monitoring.

2.2/Particulate Sampling

2.2.1 Size Measurement of Dust Particles

Principal Methods: The size of dust particles is an important parameter. The measurements of particle size distribution in particulate polluting sources are usually made for some fundamental reason such as:

1. To complete the characteristic information about a given source.
2. To permit assessment of potential health and environmental effect of different sources.
3. To obtain a complete quantification and better selection and design of control methods and devices.

It is of essential importance that the sample collected for size analysis be representative of the *entire* original material and fulfill the statistical requirement. Dust particles may be collected by means of the following collection principal methods.

2.2.1.1 Sedimentation

Particles may be collected from the atmosphere by having them deposited under their own gravity on a collecting surface. The dust samples collected by this method are by no means considered representative of airborne composition. The main reason is that particles smaller than about 5 μm stay in suspension in the atmosphere. The few that deposit out do so with a rate much less than proportional to their concentration.

Another disadvantage of this method is that the amount of sample collected cannot be related to a specific volume of air to establish the dust concentration at a given location. The sedimentation method is an economical method of assessment of atmospheric deposition analysis. The size distribution of dust particles collected by this method can be obtained by microscope examination of the deposition surface which may be a glass slide. Sieving may also be employed. Sticky surfaces are recommended for particle collection to prevent particle reentrainment after being deposited. The size analysis in this case gives the size distribution along with the number of particles per unit area (rather than mass concentration as for other conventional methods).

2.2.1.2 Filtration

While the sedimentation method of sampling fails to collect small dust particles from the atmosphere, the filtration method avoids this defficiency. Samples of atmospheric air are drawn through a filter element that retains the dust particles on its surface. The essential requirement for this method is that the filter element must be of a type that does not permit particles to penetrate deeply inside the filter body. Therefore, fibrous filters are not suitable as the particles penetrate into the filter medium and they cannot be detected under the microscope. The filter element must also have a high collection efficiency (better than 99 percent) and be capable of collecting all particles including those in the sub-micron range size. For this purpose filters having pore size from 0.01 up to 10 μm are recommended. Particles having size much less than the filter's pore size are readily collected at 99 percent efficiency due to the electrostatic charges concentrated on the filter surface. The filter must also have an acceptable resistance to the air flow.

2.2.1.3 Inertial Impingement

This method of sampling employs devices that fall into the inertial sizing category such as impactors, cyclones, centrifuges, and virtual impactors which are known as "cascade" impactors. The principle of these devices is that the air sample is constrained to follow a path of curvature so that the heavier dust particles detach themselves from the air path and move outward toward an impaction surface because of their inertia. Subsequent analysis of the particle size distribution may be made by gravimetric means, chemical analysis, or microscopic inspection. The particle size distribution is usually determined by relating the cut diameter for each stage which is previously determined by the device calibration to the mass of dust collected on these stages. The inertial impingement is a successful and widely used method of particulate sizing in the range of 0.3 to 20 μm diameter.

2.2.1.4 Electrical Precipitation

The dust sample drawn into the device is subjected to an intense electrical field. The particles become highly charged and are collected at high efficiency on a collecting surface. Electrical precipitation is highly efficient in the entire range of particle sizes. The collected particles are then examined by a microscope for size distribution determination.

Some of the principal methods for determination of the particle size distribution are given above. The suitable method must be selected according to a category that takes into consideration the objectives of the study, the working conditions, the particle size range, the type of analysis to be performed on the sample, the availability of equipment, and the economics of the study.

Among the different principal methods of particle size measurements, the inertial impingement is known to be used in most field investigations of fugitive dust. In the following sections, samplers that are based on inertial impingement and widely used will be described along with the operating procedures.

2.2.2 Determination of the Size Distribution of Dust Particles by Cascade Impactors

2.2.2.1 Principle of the Cascade Impactor

The inertia impactors use the phenomenon of the inertia dependence on the particle size at a given velocity. When the particles have inertia greater than a critical value, they can be selectively removed from an air stream at the different stages of the cascade. The cascade impactors utilize multi-stages upon which particles of different sizes are collected. Depending on the individual design, the impactors may have any number of stages. Most of the commercial impactors have 4–8 stages. The data obtained by the cascade impactor are usually expressed in terms of the aerodynamic diameter.

For the purpose of field investigation of fugitive emission or (in general) the particulate pollution, two types of cascade impactors are widely used for determination of the size distribution of particulate matter, the Andersen cascade impactor and the Research Appliance Company Fractionating Sampler (RAC). However, there are some other types of commercial instruments that utilize the same principles that can be used for the same and different purposes. The description of the Andersen and RAC impactors are given in the following sections. The procedure of operation, sampling, and analysis for both instruments are essentially identical; these will be described in Section 2.2.2.4.

2.2.2.2 Andersen Cascade Impactor Model Mark I Description

Description

1. The impactor. The Andersen cascade (see Figure 2.1) impactor consists of a series of eight stages and collection surfaces. The sample is automatically separated into eight fractions from 10.0 μm and above down to 0.4 μm diameter with a constant flow rate of 28.3 l/min (1 CFM). The eight stages of the impactor are followed by a back up filter holder. They are all held together by three spring clamps and gasket with O-ring seals. Each stage has a removable stainless steel collection plate 8.26 cm diameter (glass substrates may also be used).

The combination of a constant flow rate and successively smaller diameter orifice increases the velocity of sample air as it cascades through the impactor, resulting in the impaction of the progressively smaller particles in succeeding stages. Particle fractionation ranges from 10.0 to 0.4 μm diameter. Particles too small to be impacted on the last collection plate are collected in the backup filter which is an integral part of the sampler.

Figure 2.1. Assembled cascade impactor head (Courtesy: Anderson Samplers, Inc.).

The particle collection surfaces should have four important characteristics. They should be (1) of inert material, (2) capable of being cleaned with any solvent, (3) capable of being coated with a sticky material, and (4) light enough to be tared on a precision analytical balance. The collection surfaces must also be easily handled, cleaned, stored, and transported. Stainless and aluminum foils are typical collection surfaces that meet the above mentioned characteristic requirements.

2. Filter Unit. The eight stages of the impactor are followed by a backup filter holder which accommodate a high efficiency 81 mm filter disc. Type A glass fiber filter media or any equivalent media is used which has a high collection efficiency for submicron particles.

3. Vacuum Pump. A vacuum pump is provided with the sampler so that they are together placed in their own carrying cases. It is a rotative vane type pump capable of maintaining a constant flow of air through the impactor (28.3 l/min.) and after-filter.

4. Electrical Requirement. A line of 115/230 volts is needed 50/60 Hz AC. Also 12 Volts DC can be used for the sampler.

2.2.2.3 The RAC Fractionating Sampler

The RAC fractionating (Figure 2.2) sampler consists of five stages of a cascade impactor that is capable of separating the sample aerodynamically into five fractions from 3.0 μm down to 0.5 μm diameter at high flow rates (113–142 l/min). The unit can be used for a

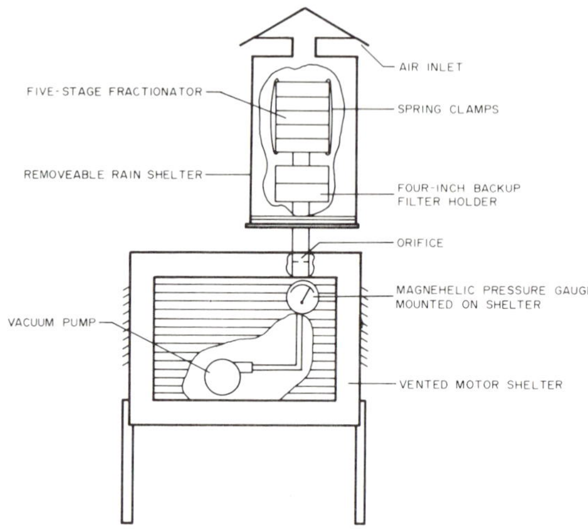

Figure 2.2. Diagram of a fractionating sampler and shelter.

24-hour sampling cycle. Figure 2.2 shows the RAC fractionating sampler in operating configuration. The main components of the sample are as follows:

1. The impactor. Consists of five in-line anodized aluminum stages. Each stage has 400 orifices (air jets) with progressively smaller cross section. The diameter of the air jets from stage 1 to stage 5 have diameters of 1.14 mm to 0.25 mm. The five stages are followed by a filter holder in which a membrane or glass fiber medium filter, having a diameter of 100 mm, is placed between two teflon gaskets. The filter holder has three bolts that must be tightened properly to secure the filter element and seal it in position. The five stages are held together with the filter holder by means of three spring clamps and gaskets with O-ring seals. Each stage has a stainless steel collection plate upon which the collection surface (aluminum foil) is to be placed. The range of particle size collected on a stage depends on the jet velocity of the stage, the jet-to-collection surface distance, and the collection characteristics of the previous stage. Particles that are not collected on the next stage where they are either impacted or pass on to the next stage until the jet velocity is sufficient for impaction.

2. Filter unit. The filter holder unit of the RAC sampler is 11.5 cm H × 15 cm diameter of anodized aluminum with a stainless steel media screen to support the filter element. The filter element is 10 cm diameter, 88 cm^2 effective area, and can be reinforced membrane or glass fiber medium, 3.0 μm minimum pore size (to prevent excessive pressure drop).

The filter holder is held to the 5 stage impactor by means of three spring clamps and gasket with O-rings. The air exit pipe from the sampler head (impactor + filter holder) is connected to the pump inlet line through a grommet on top of the pump housing and secured in mounting gasket.

3. Flow Measuring Device. The air flow through the sampler is regulated by an integral limiting orifice of 6.4 mm diameter which is monitored by a magnehelic gauge fixed on the control panel of the sampler. The magnehelic pressure gauge has a capacity of 0 to 100 cm of water. The operating range when a 3.0 μm membrane filter is used in 56 cm to 74 cm of water. Tappings across the limiting orifice are connected to two values that are operated by means of two buttons beside the magnehelic gauge. By pushing the first button, the tappings are connected to the gauge to read the pressure drop whenever needed during operation, the other button disconnects the tappings from the gauge so that the pointer of the gauge is released and return to zero value.

4. Vacuum Pump. The air is drawn through the system by means of a ½ hp, 200 l/min free flow, continuous oil-less, carbon vane vacuum pump. The pump with its driving motor is placed in their ventilated compartment.

5. Electrical Requirement. The RAC sample works on a line voltage 115/230 volts, 60 Hz; 7.8 amp with 115 V or 3.9 amp with 230 volt.

2.2.2.4 Operation Procedure for the Cascade Impactors

This section describes, in a general manner, the operation procedures that are essentially identical for both the Andersen and RAC impactors. However, if a specific operation procedure is needed, it will be referred to in the context. The operation is divided into three categories, namely, preparation, sampling, and analysis.

I. Preparation

1. Cleaning the impactor is considered to be one of the critical preparation steps. The impactor must be disassembled to clean the nozzles, the collection surfaces and the filter holder. Special care should be taken to clean the impactor nozzles. If they are fairly clean, a wiper such as a lint-free tissue dipped in acetone, or other suitable solvent, and tweezers can be used to reach into corners. Inspection under the microscope is sometimes useful to detect if some nozzles are partially clogged. If, however, the nozzles cannot be cleaned by applying solvent, compressed air can help in pushing dirt out of the nozzles. For strongly adhered dirt that is difficult to be dislodged by the previous

steps, the nozzle may be placed in an ultra-sonic bath for a short time (about 20 seconds). The nozzles must be carefully inspected after this bath to check against any possible damage.

2. The collection surfaces (aluminum disk, cellulose, glass fiber, etc.) as well as the filter element must be inspected to be sure that they are all clean and undamaged. The collection surface and the filter must then be marked for future identification. If necessary (especially for the aluminum foils and the cellulose), a very thin layer (fingerprint thickness) of grease or pure vaseline should be applied.
3. The collection surfaces and the filter(s) are then desiccated in the balance room for at least 24 hours to equilibrate before weighing them.
4. Use a semi-micro analytical balance to determine the weight of the collection surfaces and the filter(s). The balance must have a sensitivity of at least 0.01 mg. It is preferable to weigh them twice and the average weight is then recorded.
5. The impactor is now to be assembled. This must be done in a clean air environment (usually in the laboratory and not in open areas or at the field). Use the tweezers to mount the collection surfaces on the collection plates and assemble the stages in the proper order. Check that the O-rings are properly seated in their grooves. The filter is to be placed on the supporting screen inside the holder with the teflon gasket in place. The filter holder is then closed, tightened, and held together with the impactor by the spring clamps.
6. The impactor is then connected to the pump in a clean environment and operated briefly in order to check that the flow rate is within the working limit, there are no leaks, and the unit is working satisfactorily. After this, disconnect the impactor from the pump and seal the inlet port and the filter outlet. The whole unit is now transported to the site for sampling.

Sampling

1. Unseal the impactor inlet port and the after filter outlet and then connect the impactor to the vacuum pump and set up the unit in the sampling location. Orient the rain shelter that is usually provided to protect the impactor so that rain or other large bodies do not enter the impactor inlet. If insects, leaves, or large lint are a problem, a medium mesh screen should be used at the sampling inlet. The screen area should be larger in comparison to the sampling inlet area. If the inlet air is to be ducted, a short duct must be used to avoid the loss of dust particles by deposition on the interior surface of the duct.
2. Start sampling by switching the pump on at the desired time and record the flow rate as well as the sampling location along with any other specific information for future reference.

3. After the desired sampling time is elapsed, the flow rate and the necessary information are recorded, then stop the sampler.
4. Disconnect the impactor and filter holder assembly, seal off the impactor inlet, and the after filter outlet to prevent dust particles from penetration inside the impactor during transportation to the laboratory.

Analysis

1. Clean the external surfaces of the impactor and filter holder assembly before disassembling the unit to avoid the contamination of the collection surfaces or the filter upon removal. Disassemble the consecutive stages of the impactor and remove the collection surfaces and then the filter element and place all in a drying oven with a maximum temperature of about 40 °C. A desiccator may be used for this purpose in which the samples must stay for 24 hours before weighing. Upon removal of the collection surface, note their identification numbers and positions.
2. After the drying process, weigh the collection surfaces and the filters to the nearest 0.01 mg and record the weights along with their identification numbers. The weights of the collected dust particles are then determined by subtracting the final, minus the initial weights. The collected materials may be analyzed chemically after the weighing process.

2.2.2.5 Data Use Example

The following data are from Dr. Michael J. Pilot, developer of the University of Washington Cascade Impactor. They serve to show a simple procedure for the use of impactor data to determine particle size and distribution. For the seven impactor stages and the back up filter, the following aerodynamic cut diameter (d_{a50}) and *dry* sample weights are obtained:

Stage #	d_{a50}, μm	Sample Mass, g
1	29.15	.01320
2	10.50	.00830
3	4.73	.00890
4	2.35	.00590
5	1.29	.00490
6	0.65	.00230
7	0.34	.00170
Filter Mass		.00910
Total Mass		.05430

First, add the sample mass to obtain the total noted. Then divide each mass by the total and convert to percent to obtain sample percent by mass in each size range as shown below:

d_{a50}, μm	Sample % by Mass in Size Range	Cum % by Mass Less Than Stated Size
29.15	24.31	75.70
10.50	15.29	60.41
4.73	16.39	44.02
2.35	10.87	33.15
1.29	9.02	24.13
0.65	4.24	19.89
0.34	3.13	16.76
Filter	16.76	—

The final column is the summation of the mass of material less than the stated size. When this column is plotted versus d_{a50} on a log probability plot (such as Figure 1.2), the mass mean diameter is found to be 6.3 μm and the particle distribution is bimodal (two straight, intersecting lines representing two or more basic materials).

2.2.3 Calibration of Cascade Impactors

Basic Requirements

1. It is extremely important to know the collection characteristics of the cascade impactor being used in a measurement, i.e., the measurable size range and the cut diameter for each stage at different flow rates. The sampled particles are usually polydispersed (different sizes), with various components of different density as well as nonspherical shapes, so that accurate calibration of a cascade impactor is considered to be a difficult job. All cascade impactors used in particulate investigation work on similar principles and have almost identical design. Therefore, calibration procedure for most impactors are the same.
2. The aerodynamic diameter of particles collected on each stage of an impactor depends mainly on the nozzle diameter and flow rate in the stage. It is, therefore, of great importance to first ensure the cleanliness of the nozzles before calibrating the impactor and checking them for any damage. Second, the impactor must be inspected for possible air leaks, and must be eliminated completely whenever

found. Air leaks increase the true sample flow rate and consequently the collection characteristics of the impactor in an uncontrolled manner. Air leaks also increase the internal particle loss, i.e., the amount of particles that are deposited on internal parts of the impactor other than the collection surfaces.

3. Should a sticky material be applied on the collection surfaces to ensure particle adhesion, it must meet the following important requirements: (a) weight stability, (b) uniformly distributed over the surface, (c) non-reactive with the deposited particles, (d) non-volatile after being weighed, (e) non-hygroscopic.
4. Precautions must be taken to prevent particle bounce after being deposited on the collection surface. Bounce may occur depending on the nature of the collected particle. For example, mineral particles are most likely to bounce while non-volatile oil droplets are not. Bounce also may occur for particles impacted on aggregate of deposited particles. Therefore, a larger collection surface can accommodate larger amounts of particles without forming aggregates and with less bounce.

Calibration Method

1. The first requirement is to have standard particles of uniform size (1) that can be used to calibrate the individual stages of the impactor. The calibration must be carried out using an accepted technique [2]. Calibration data must include the following: (a) the dimensions of each impactor nozzle, (b) the collection surface size and the coating used, (c) the air flow rate, (d) the calibration particle size, shape, density, and method of generation, (e) a method of particle size measurement and the estimated accuracy, (f) the method of measuring the particle weight and the estimated particle loss within the impactor.
2. Three calibration tests must be performed for each impactor stage. During these tests the flow rate and the particle size, density, and shape must be fixed. The entire cascade impactor should be calibrated using test particles in the range of 0.30 to about 20 μm diameter. The average collection efficiency of at least two test runs must then be used for plotting the test data. As a guideline, one test condition must result in a stage collection efficiency between 50 and 80 percent, for the second test condition between 20 and 50 percent.
3. The collection efficiency is then plotted versus the inertial impaction parameter on a log-probability plot. A best fitting straight line should be drawn through at least three test condition points. The resultant plot can be used to determine the impaction parameter value for the stage 50 percent collection efficiency. The advantage of using the impaction parameter is that it allows calibration tests to be

done at flow rates different from the design flow rate. However, the flow rate must not be less than half or higher than twice the design flow rate. Figure 2.3 shows the plot of impactor stage collection efficiency versus the inertial impaction parameter, K. The impaction parameter is defined as:

$$K = \left(\frac{C \varrho_p V_j d_p^2}{18 \mu d_j} \right)^{1/2}$$

where C is the cunningham correction factor (section 1.2.2)

ϱ_p = the particle density g/cm³
d_p = the particle diameter cm
d_j = the jet diameter (or width) cm
V_j = average velocity at the impactor jet exit cm/s
μ = air viscosity, poise

4. The impactor stage particle cut size can be determined from Figure 2.3 which corresponds to a 50 percent efficiency value on the calibration data plot at the design flow rate.
5. The test particles must be of spherical shape with a density determined to have a maximum standard deviation of 10 percent. The standard

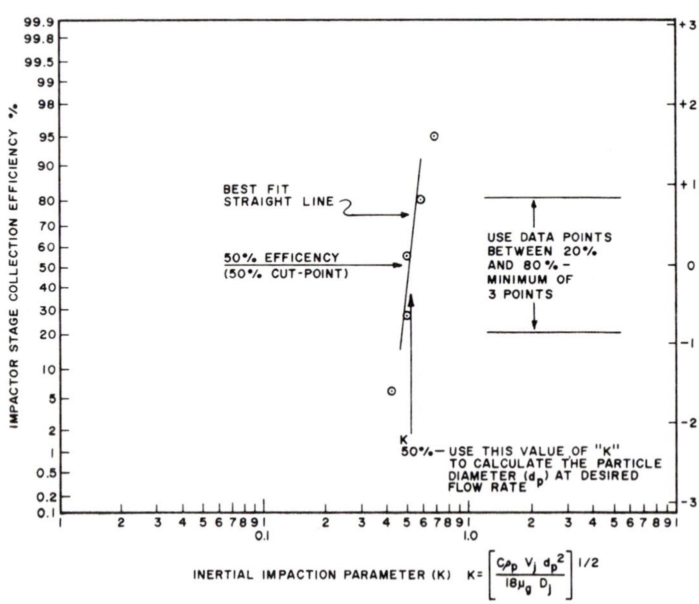

Figure 2.3. Typical impactor stage calibration data of stage efficiency versus inertial impaction parameter.

deviation of the particle size must not exceed 10 percent. To meet these requirements, particles may be generated by one of the standard methods [3,4] that are capable of generating uniform aerosol particles. The particle size can be determined by means of optical microscopes.

2.3/The High-Volume Air Sampler

2.3.1 Main Features

The high-volume sampler is based on gravimetric principles and capable of sampling large volumes of air for the collection of suspended particulate matter. A known volume of air is drawn through a glass fiber filter of known weight. Reweighing of the filter after sampling under controlled conditions gives a direct measurement of particulate mass. Particles as small as 0.01 μm diameter can be collected by this sampler. The high volume is designed and built for continuous 24 hour a day sampling and is provided with a heavy duty turbine blower at a constant flow rate of from 39 to 60 cfm (1.1 to 1.7 m^3/min.). During the sampling operation, the sampler is supported in a protective shelter and the filter is in a horizontal position. Collaborative testing for similar gravimetric methods showed that the coefficient of variation for single operator variation (repeatability of method) is 3 percent [5]. The corresponding value for multi-operator variation (reproducibility of the method) is 3.7 percent.

2.3.2 Sensitivity of Operation and Accuracy

With the high volume sampling at an average flow rate of 60 cfm (1.7 m^3/min) over 24 hours, a sufficient measurable sample will be obtained even in an atmosphere a particulate concentration of 1 $\mu g/m^3$. However with high particulate loaded atmosphere, a shorter sampling time should be adopted. Preliminary tests and skill of operation provide better estimation of the sampling time. The collected particulate matter is determined to the nearest milligram and air flow rate to the nearest 0.03 m^3/min.

The standard deviations of the measured concentration by means of the high-volume were found [6] to be 10 $\mu g/m^3$ for an average mass concentration of 112 $\mu g/m^3$ and 6 $\mu g/m^3$ for an average mass concentration of 39 $\mu g/m^3$.

The accuracy of concentration measurements depends primarily upon the uniformity of the air flow rate, the type of sampler, and possibly the sampler orientation. With a progressive accumulation of particulate on the filter and depending on the nature of dust, the air flow rate error may be significantly reduced. Errors associated with a drastic reduction in flow rate may be as high as ±50 percent.

2.3.3 Limitations

For the cases in which oily particulates exist such as photochemical smog or wood smoke, the use of the high-volume sampler may clog the filter and cause a rapid drop in air flow at a non-uniform rate. High humidity and dense fog can also lead to the same result.

While the glass-fiber filters are comparatively insensitive to change in relative humidity, the collected particulate can be hygroscopic. It is, therefore, necessary to consider carefully these phenomenon during the operation.

2.3.4 Description of the Apparatus

The high volume sampler consists mainly of the following parts (see Figure 2.4):

1. The Sampler. The sampler consists of a blower unit to which a filter holder is attached. The filter holder (Figure 2.5) consists of two parts: a) A stainless steel filter adapter (20 × 25 cm) with a rectangular opening at the top, covered with a screen which ends in a circular screw on a connector at the bottom; and b) an open rectangular face plate of aluminum with a sponge rubber gasket.

In sampling, a filter is placed between the filter support screen and the gasket face plate. The power requirement for the sampler is approximately 550W, 110V (AC) and 7A.

2. Flow System. This provides a continuous flow measurement and

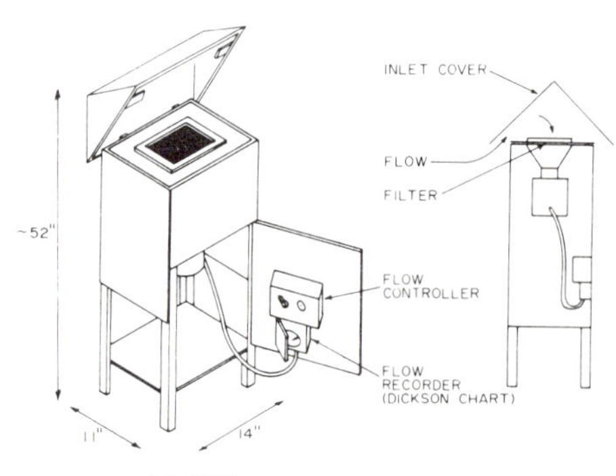

Figure 2.4. High volume sampler "dog house" type.

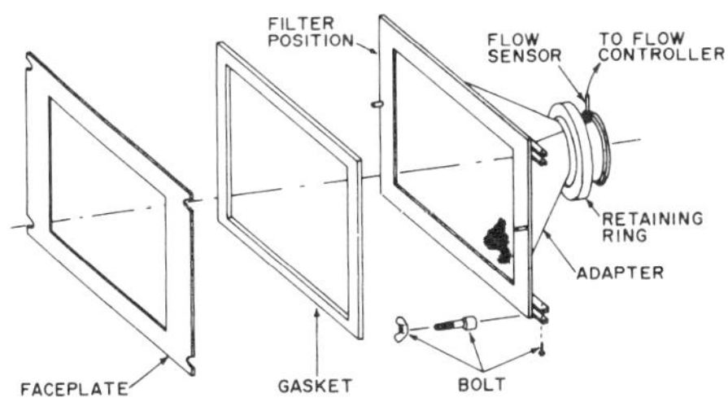

Figure 2.5. High volume filter holder assembly.

recording, by using a "mini-corder" which is installed on the front of the sampler shelter. The recorder is connected to the blower housing by means of a Tygon tube and runs continuously, independently of the sampler. On start up, the pen rises from zero making a trace that indicates the start time and then points to the actual flow rate. The chart of the recorder will show the start time, stop time, the duration of power interruption, if any, as well as the flow rate during the operating period.

3. Flow Controller. It is a device that is capable of controlling the flow through the sampler to $\pm$ 1 ft.^{3}/min. (0.028 m^3/min) under variations in filter loading.

4. Time Measurement. An electromechanical time meter is located on the face of the flow controller to measure the total sampling time in minutes and can be reset by pushing a button to return all counters to zero. It is turned off and on simultaneously with the sampler by the master timer.

5. Master Timer. It controls the timed operation high-volume sampler. It should be checked at each sample change to make sure that the next sampling period will be correct. This is shown in Figure 2.6.

6. Sample Shelter. This protects the sampler against the extremes of environmental conditions such as high and low temperatures, high humidity, rain, and any type of air pollutants. It is usually aluminum, but if it is to be protected from rusting, apply a heavy duty exterior paint to the shelter. The clearance area between the edge of the roof and the main housing should be 650 cm^2 $\pm$ 50 cm^2, and the main housing should be rectangular having dimensions of 30 $\times$ 35.5 cm.

Usually supplemental devices are secured to the shelter. These include: differential manometer (at least 40 cm of water); rotameter; positive displacement meter (to be used as a certified primary standard); and a barometer for measuring the atmospheric pressure for the orifice calibration unit.

7. The Orifice Calibration Unit (Figure 2.7). It consists of a metal tube 7.62 cm (3″) I.D. and 15.9 cm (6¼″) long with a static pressure tap 5.08 cm (2″) from one end. The tube end nearest the pressure tap is flanged to

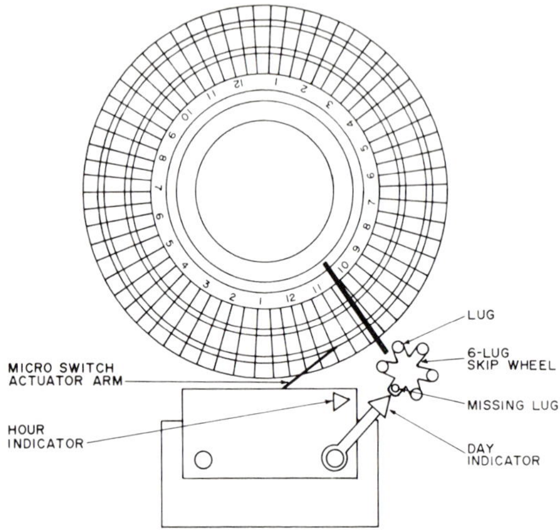

Figure 2.6. Tork high volume master timer.

about 10.8 cm (4¼″) O.D. with a male thread of the same size as the inlet end of the sampler. A single metal plate 9.2 cm (3 5/8″) in diameter and 0.24 cm (3/32″) thick with a central orifice 2.86 cm (1 1/8″) in diameter is held in place at the air inlet end with a male threaded ring. The other end of the tube is flanged to hold a loose female threaded coupling, which screws on to the inlet of the sampler. An 18-hole metal plate (or a 13, 10, 7, or 5 hole plate), an integral part of the unit, is positioned between the orifice and sampler to simulate the resistance of a clean filter (see Figure 2.8).

2.3.5 Filter Quality

The recommended filter media to be used with the high-value sampler is glass-fiber filters with a collection efficiency of at least 99 percent for particles 0.30 μm diameter. However, some other medium may be desirable for some analysis, such as paper. Filters that contain high background concentration of the pollutant being investigated must not be used.

2.3.6 Operation Procedure of the High-Volume Sampler

1. Filter Preparation. Filter paper elements must be numbered using a numbering device. The filters are then placed in a drying oven at a temperature of about 45 °C for a period of at least 10 hours. A desiccator must be used to store the filter which must be equilibrated to the condi-

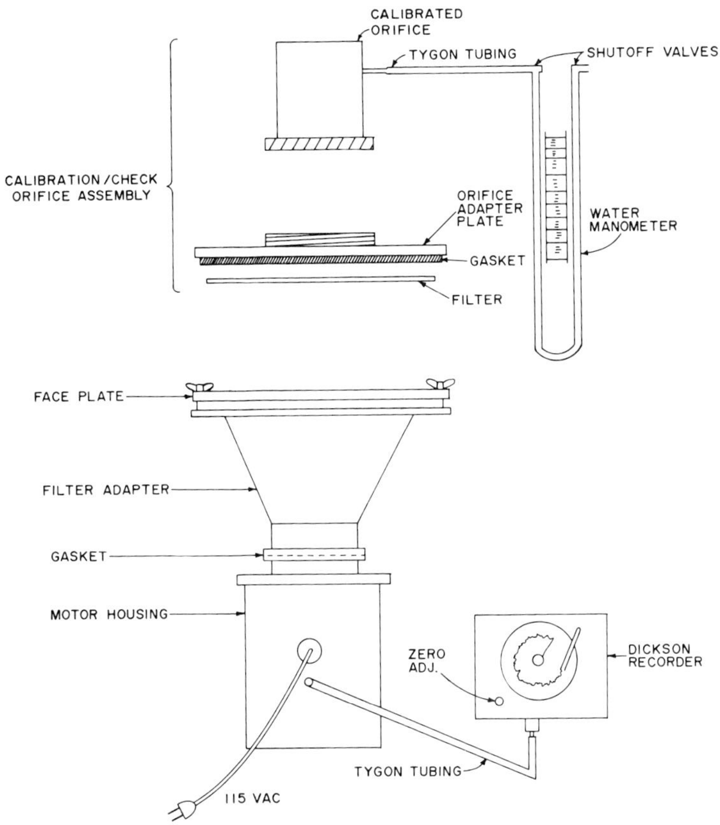

Figure 2.7. High volume orifice calibration unit.

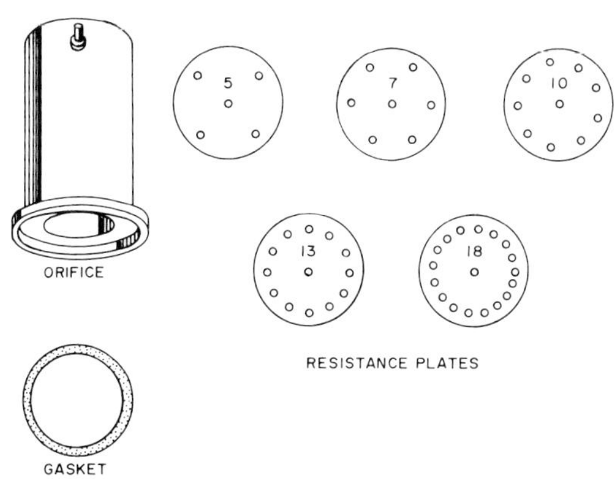

Figure 2.8. Resistance (restrictor) plates for orifice calibration unit.

tions of the balance room for at least 24 hours. The filters are then weighed using an analytical balance that must be equipped with a weighing chamber designed to handle unfolded filters and having a sensitivity of 0.1 mg. The filters must not be bent or folded before sampling. They may be stored in a dry and clean place for future use.

2. Sampling Operation. In the light of the current experiences, the following operational steps are recommended to obtain reliable results [7].

a. Examine the screen. If it is dirty, it should be wiped clean with a clean paper towel. If it is damaged, fix or relace it. Then place a clean filter in positive on the screen of the filter holder. Fiberglass filters have a smooth and rough side. The smooth side should be placed *down* so the rough side is *toward* the sample. Be sure the filter is centered on the screen so that when the face plate is in position, the gasket will make an airtight seal on the outer edges of the filter.
b. Place the plate in position on the filter holder being careful not to move the filter out of position. Move the bolts into place and gently tighten the wing nuts, working from opposite corners. It is important that the wing nuts be tightened evenly and uniformly to prevent air leakage around the filter. If they are too tight, the gasket may never recover its elasticity sufficiently to seal properly. On the other hand, if the plate has not been sufficiently tightened, the edges of the sample area will be irregular and signs of air leakage will be shown by streaks across the otherwise clean margins of the filter paper. An aluminum cover is used to cover the filter if the sampling will not start immediately to protect the filter from any possible contamination. Just before sampling, uncover the filter. Close the roof of the shelter carefully to avoid damaging the filter.
c. Install a new chart on the pressure recorder. Record sampler type, site, filter number, and sampling date on the back of the chart before placing on the recorder. Set the chart at the proper starting position and ensure that the edges are properly located under the two small retainers.
d. Advance to correct starting time. Turn the drive spindle clockwise to the required time.
e. Zero the pen by gently tapping the side of the recorder and adjusting the potentiometer, if necessary.
f. At the end of the sampling period, open the roof of the shelter and place the aluminum cover plate over the filter. Unscrew the four wing nuts, holding the face plate until the bolts can be pushed back sufficiently to permit the removal of the entire plate and filter assembly. Remove the face plate by lifting it up carefully.

g. Uncover the filter and with great care, use the corner of the filter holder to lift the filter from the holder. Slide the folder under the filter, center it, and fold carefully at the center of the exposed area making sure that exposed areas overlap and do not contact the clean filter margins. Examination of the filter at the end of a sampling period will show whether the filter was properly placed and sealed. The edges of the sample area should be sharply defined with a 1.25 cm clean margin on each side of the filter paper.
h. Place the folder containing the filter in a prepared sheet, seal, and transfer to the laboratory for analysis.
i. Remove the chart and be sure to record all the necessary information on the back of the chart.
j. Record the field data on the data sheet.

2.3.7 Validation Criteria of Sampling Process

In order to assist the operator in obtaining valid results and better performance of the high-volume sampler, the following precautions should be followed:

1. Under adverse weather conditions, precautions must be taken to avoid damage of the filter. During the periods of high wind or heavy precipitation, it may be necessary to turn off the sampler and postpone removal of the filter until weather conditions improve.
2. Sometimes the filter adheres to the gasket when the face plate is removed. The filter may be dislodged by gently jarring the face plate. Applying a small amount of talc powder on the gasket and rubbing it in before installing new filter eliminates such a problem. Avoid overtightening the wing nuts.
3. If the filter becomes wet during a severe storm, the motor may overheat sufficiently to damage it permanently. It is, therefore, advisable to shut off the sampler during severe storms, if possible.
4. Decreases in flow rate during sampling of more than 10 percent from the initial set point invalidate the test. Changes in flow rate calibration are determined by a field calibration check.

2.3.8 Calibration of the High-Volume

The operator must be aware of the accurate performance of the sampler and to be sure that the percent error in actual flow rate is kept within the allowable limit (±10 percent). The accuracy of the flow rate is usually checked through a regular calibration of the instrument. Two calibration routines are necessary for the high-volume sampler. The first, a one point field calibration check should be done every other sampling period. This is to determine whether the percent error is greater than ±10 percent. If the percent error is greater than ±10 percent, the second calibration is necessary. This is the five-point calibration. The five-point calibration may be carried out with the High Volume equipped with a chart recorder or with a rotameter for flow measurements.

2.3.8.1 One-point Field Calibration

The master timer must be bypassed in this case in order to keep the sampler working continuously. To bypass the timer (see Figure 2.6):

1. Rotate the skip wheel until the day indicator is pointing to the sampling day (at the missing lug).
2. Once the wheel reaches this position, the power is supplied to the sampler. When the calibration is completed the power is disconnected by rotating the skip wheel and the timer can be adjusted as follows:
   a. Set the hour dial to the present time.
   b. Move the skip wheel clockwise and set it so that the number of lugs between the missing lug and the day indicator is equal to the number of days before the next sampling date.

Also required for calibration are:

Calibration Kit (Figure 2.9)
Orifice Curve (Figure 2.10)
Sample Interpolation Table (Table 2.1)
Hi Vol Calibration Curve (Figure 2.11)
Flow Rate Calibration—Interpolation Table (Table 2.2)

The calibration procedure is summarized in step-wise fashion in Table 2.3 and an example of this follows.

Example: In a one point-field calibration process for a high volume sample, the Dickson chart recorder's reading is 40. The manometer reads a differential pressure head through the calibration orifice of 7.9 in water. Calculate the air flow rate and the percent error in the observed recorder reading.

Solution: Using Figure 2.7 and Table 2.2: For Dickson reading 41, the flow rate = 1.446 m^3/min.

Using Figure 2.7 and Table 2.1: For an orifice manometer reading of 7.9 inches water, the air flow rate = 1.394 m^3/min.

$$\text{The percent error} = \frac{1.446 - 1.394}{1.394} \times 100 = 3.7$$

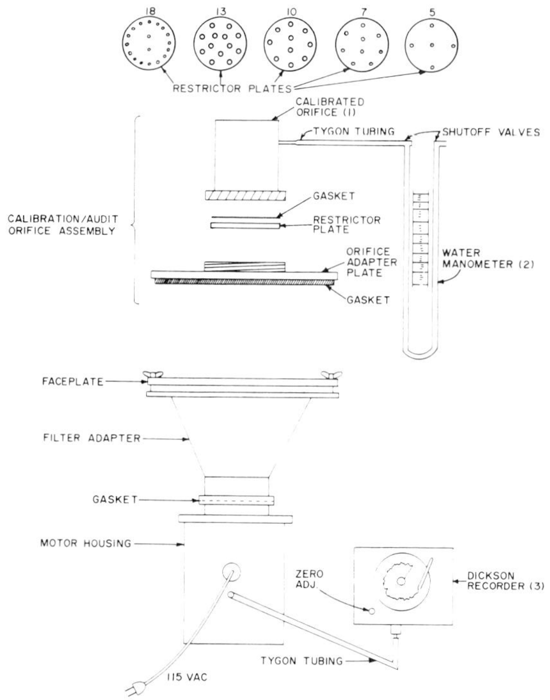

Figure 2.9. High volume calibration kit.

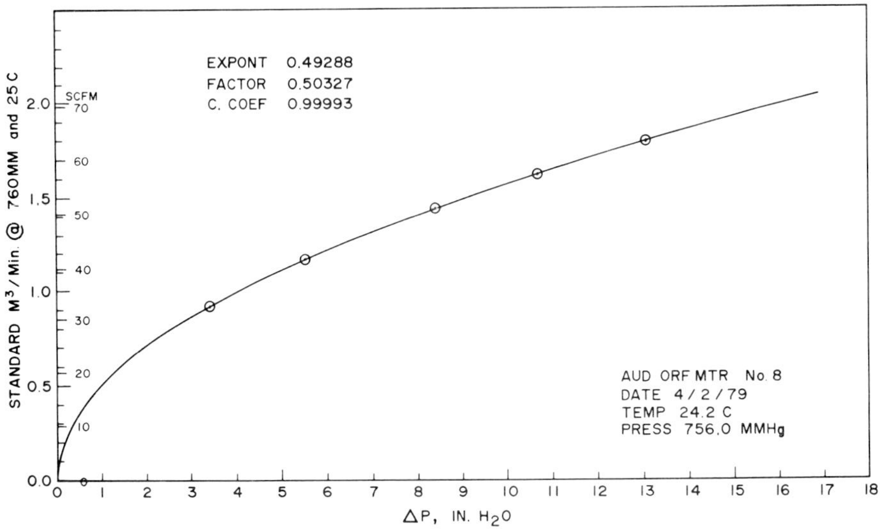

Figure 2.10. Sample high volume flow orifice calibration curve from the inhalable particulate network.

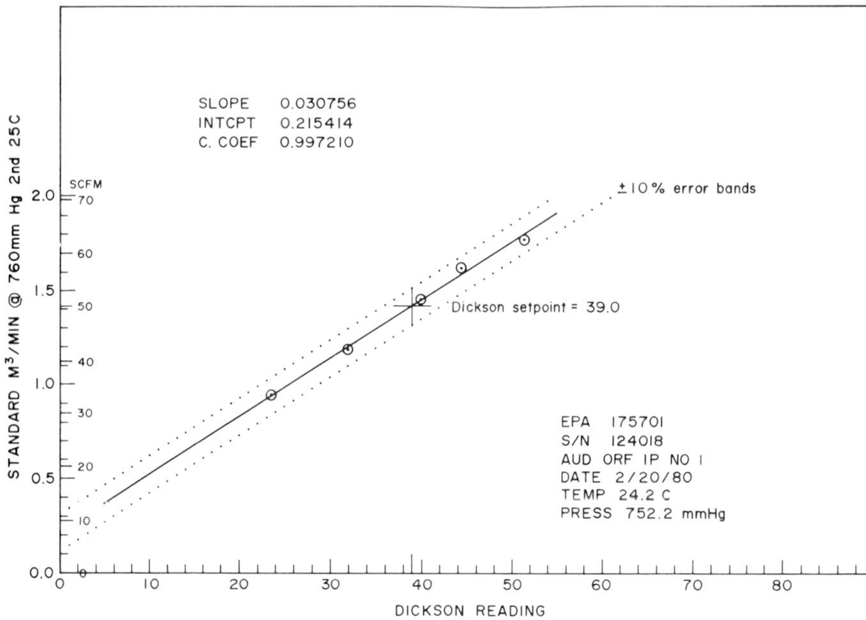

Figure 2.11. Sample high volume sampler flow rate calibration curve from the inhalable particulate network.

A percent error of ±10 would be acceptable for the high volume. If this calibration shows higher values, the five-point calibration is necessary.

2.3.8.2 Five-point Calibration

This calibration should be made and can be performed with either the air flow recorder or a rotameter. The calibration kit for use with the recorder is shown in Figure 2.9 (including the five restriction plates). In some cases the high volume sampler is not provided with an air flow chart recorder, a rotameter may be used for air flow reading. In such a case, the method of calibration is changed very slightly from the previous five-point calibration method.

The two calibration methods are essentially the same, except that with every calibration restriction plate in place, and with the sampler running, the associated rotameter and manometer readings are recorded. Convert the differential manometer reading to m^3/min., Q, then plot rotameter reading versus Q. Identify this plot with the particular sampler and rotameter combination used in its preparation. The step-by-step calibration procedure is given in Table 2.4 for the chart recorder.

Table 2.1. Sample Interpolation Table for High Volume Flow Orifice Calibration.

MAN RDG	SCFM	M^3/MIN	MAN RDG	SCFM	M^3/MIN	MAN RDG	SCFM	M^3/MIN
2.0	25.008	0.708	5.0	39.284	1.113	8.0	49.524	1.403
2.1	25.617	0.725	5.1	39.669	1.123	8.1	49.829	1.411
2.2	26.211	0.742	5.2	40.051	1.134	8.2	50.131	1.420
2.3	26.791	0.759	5.3	40.428	1.145	8.3	50.431	1.428
2.4	27.359	0.775	5.4	40.803	1.156	8.4	50.730	1.437
2.5	27.915	0.791	5.5	41.173	1.166	8.5	51.027	1.445
2.6	28.460	0.806	5.6	41.540	1.176	8.6	51.322	1.453
2.7	28.994	0.821	5.7	41.904	1.187	8.7	51.615	1.462
2.8	29.519	0.836	5.8	42.265	1.197	8.8	51.906	1.470
2.9	30.034	0.851	5.9	42.623	1.207	8.9	52.196	1.478
3.0	30.540	0.865	6.0	42.977	1.217	9.0	52.485	1.486
3.1	31.038	0.879	6.1	43.329	1.227	9.1	52.771	1.494
3.2	31.527	0.893	6.2	43.678	1.237	9.2	53.056	1.503
3.3	32.009	0.906	6.3	44.023	1.247	9.3	53.340	1.511
3.4	32.483	0.920	6.4	44.366	1.256	9.4	53.622	1.519
3.5	32.951	0.933	6.5	44.707	1.266	9.5	53.902	1.527
3.6	33.411	0.946	6.6	45.044	1.276	9.6	54.181	1.534
3.7	33.866	0.959	6.7	45.380	1.285	9.7	54.458	1.542
3.8	34.314	0.972	6.8	45.712	1.295	9.8	54.734	1.550
3.9	34.756	0.984	6.9	46.042	1.304	9.9	55.009	1.558
4.0	35.192	0.997	7.0	46.370	1.313	10.0	55.282	1.566
4.1	35.623	1.009	7.1	46.695	1.322	10.1	55.554	1.573
4.2	36.049	1.021	7.2	47.018	1.332	10.2	55.824	1.581
4.3	36.469	1.033	7.3	47.339	1.341	10.3	56.093	1.589
4.4	36.885	1.045	7.4	47.658	1.350	10.4	56.361	1.596
4.5	37.296	1.056	7.5	47.974	1.359	10.5	56.628	1.604
4.6	37.702	1.068	7.6	48.288	1.368	10.6	56.893	1.611
4.7	38.104	1.079	7.7	48.600	1.376	10.7	57.157	1.619
4.8	38.501	1.090	7.8	48.910	1.385	10.8	57.419	1.626
4.9	38.895	1.101	7.9	49.218	1.394	10.9	57.681	1.634

M^3/MIN = 0.503267 (Manometer reading); exponent value of 0.492880
SCFM = 17.770713 (Manometer reading); exponent value of 0.492880
Manometer reading = manometer reading, in. H_2O
M^3/Min = cubic meters/min (25C,760 MMHg)
SCFM = cubic feet min (25C,760 MMHg)

2.3.9 The High-Volume Sampler with Size Selective Inlet

Fugitive dust in the size range of 0 to 15 μm diameter is of principle concern in problems of particulate pollution. Sampling of dust in this size range at high rate can be made by using a high volume sampler that is capable of fractionating the dust and allowing only the size of concern (0–15 μm) to be collected on the filter element. This high volume sampler is essentially identical to the sampler described in Section 2.3.4., except that the shelter and the face plate are removed and replaced by an inlet fractionator. A description of the sampler is given herewith below.

Table 2.2. Sample TSP High Volume Sampler Flow Rate Calibration—Interpolation Table.

Dixon RDG	SCFM	M^3/Min	Dixon RDG	SCFM	M^3/Min	Dixon RDG	SCFM	M^3/Min
25.0	34.757	0.984	40.0	51.047	1.446	55.0	67.338	1.907
25.5	35.300	1.000	40.5	51.590	1.461	55.5	67.881	1.922
26.0	35.843	1.015	41.0	52.133	1.476	56.0	68.424	1.938
26.5	36.386	1.030	41.5	52.676	1.492	56.5	68.967	1.953
27.0	36.929	1.046	42.0	53.219	1.507	57.0	69.510	1.969
27.5	37.472	1.061	42.5	53.762	1.523	57.5	70.053	1.984
28.0	38.015	1.077	43.0	54.305	1.538	58.0	70.596	1.999
28.5	38.558	1.092	43.5	54.848	1.553	58.5	71.139	2.015
29.0	39.101	1.107	44.0	55.391	1.569	59.0	71.682	2.030
29.5	39.644	1.123	44.5	55.934	1.584	59.5	72.225	2.045
30.0	40.187	1.138	45.0	56.477	1.599	60.0	72.768	2.061
30.5	40.730	1.153	45.5	57.020	1.615	60.5	73.311	2.076
31.0	41.273	1.169	46.0	57.563	1.630	61.0	73.854	2.092
31.5	41.816	1.184	46.5	58.106	1.646	61.5	74.397	2.107
32.0	42.359	1.200	47.0	58.649	1.661	62.0	74.940	2.122
32.5	42.902	1.215	47.5	59.193	1.676	62.5	75.483	2.138
33.0	43.445	1.230	48.0	59.736	1.692	63.0	76.026	2.153
33.5	43.988	1.246	48.5	60.279	1.707	63.5	76.569	2.168
34.0	44.531	1.261	49.0	60.822	1.722	64.0	77.112	2.184
34.5	45.074	1.277	49.5	61.365	1.738	64.5	77.655	2.199
35.0	45.617	1.292	50.0	61.908	1.753	65.0	78.198	2.215
35.5	46.160	1.307	50.5	62.451	1.769	65.5	78.741	2.230
36.0	46.703	1.323	51.0	62.994	1.784	66.0	79.284	2.245
36.5	47.246	1.338	51.5	63.537	1.799	66.5	79.827	2.261
37.0	47.789	1.353	52.0	64.080	1.815	67.0	80.370	2.276
37.5	48.332	1.369	52.5	64.623	1.830	67.5	80.913	2.291
38.0	48.875	1.384	53.0	65.166	1.845	68.0	81.456	2.307
38.5	49.418	1.400	53.5	65.709	1.861	68.5	81.999	2.322
39.0	49.961	1.415	54.0	66.252	1.876	69.0	82.542	2.338
39.5	50.504	1.430	54.5	66.795	1.892	69.5	83.085	2.353

M^3/Min = 0.030756 (Dixon reading) + 0.215414
F^3/Min = 1.086023 (Dixon reading) + 7.606414
Dickson setpoint for 1.42 m^3/min (50 cfm) = 39.0
Flowcheck ± 10% Dickson reading range = 35.1 to 42.9
Audit orifice = IP 1 Cal date =11/21/79
factor =0.502520 exponent = 0.424080
Dixon RDG = Dickson Reading
M^3/Min = Cubic meters/min
F^3/Min = Cubic Feet

2.3.9.1 Description of the Size-Selective Inlet (SSI) High-Volume Sampler

The high-volume sampler with inlet fractionator is a conventional TSP, equipped with an inlet section to collect particles with aerodynamic diameter in the range of 0 to 15 μm. The nominal operating

Table 2.3. One-Point High-Volume Calibration Procedure.

Step	Procedure
1	Check all fittings for leaks.
2	Remove the face plate and attach the orifice calibration unit (Figure 2.9).
3	Insert fresh filter between the orifice calibration unit and the filter adapter.
4	Connect one end of the differential manometer to the pressure tap of the calibrated orifice, the other end is open to the atmoshere.
5	With the motor power cord connected to the flow controller, operate the sampler using the master timer.
6	After one minute record the Dickson Reading.
7	Use the calibration curve 2.11 and interpolation Table 2.2 to obtain the sampler flow rate (m^3/min).
8	Read the orifice rate corresonding to the orifice Δp (inches of water), Figure 2.10 and Table 2.1.
9	Advance the recorder chart with a screwdriver to make a trace on the chart.
10	Switch off the sampler at the master timer.
11	Calculate the percent error in the flow rate as described in the following example.

Table 2.4. Five-Point Calibration of High Volume With Air Flow Recorder.

Step	Procedure
1	Check all fittings for leaks.
2	Replace the face plate of the sampler by the calibration orifice.
3	Load No. 106 chart on the recorder and set it to zero.
4	Connect the manometer to the pressure top of the calibrated orifice.
5	Without using a filter, insert the 18-hole restriction plate between the calibration orifice and the orifice adapter and tighten securely.
6	Disconnect the motor power of the sampler from the flow controller.
7	Connect the motor to a power supply and keep it running for one minute.
8	Record the recorder, manometer readings (inches water) and the plate number.
9	Advance the recorder chart using a screwdriver and write the plate number adjacent to the trace.
10	Repeat step 7 through 10 for the 13, 10, 7, and 5-hole resistance plates. This completes the calibration.
11	Replace the faceplate by the calibration orifice and remove the chart.
12	Convert manometer reading to m^3/min, Q, then plot with particular sampler and recorder used in its operation.

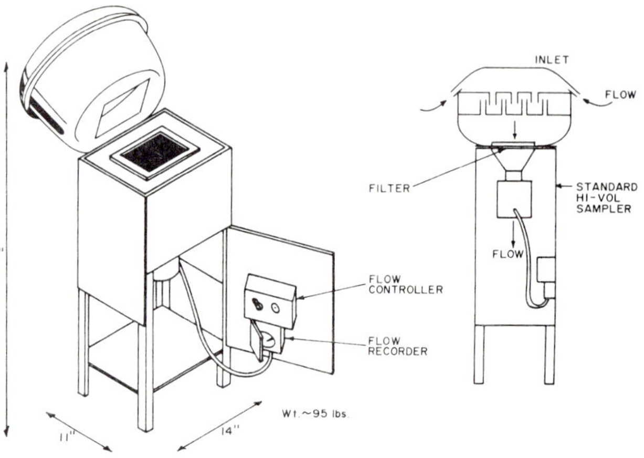

Figure 2.12. High volume sampler with size selective inlet.

flow rate is about 20 percent less than that of the TSP high volume sample, it is 1.13 m^3/min. (40 ft^3/min.) compared with 1.42 m^3/min. (50 ft^3/min.). The major changes from the TSP high volume are the removal of the face plate and hold-down screws; these are replaced by a spring clamp mechanism. Figure 2.12 shows the high volume sampler with the size-selection inlet on the top, and Figure 2.13 shows a cross section in the selection inlet. The size fractionator consists mainly of specially designed baffles which are arranged to form acceleration nozzles. As the dust passes through these nozzles, particles larger than 15 μm aerodynamic diameter have a momentum sufficient to separate them from the main flow and cause them to impact on the wall of the chamber where they are collected. The rest of the dust particles ($\leqslant 15$ μm) continue to flow to the filter and are collected there.

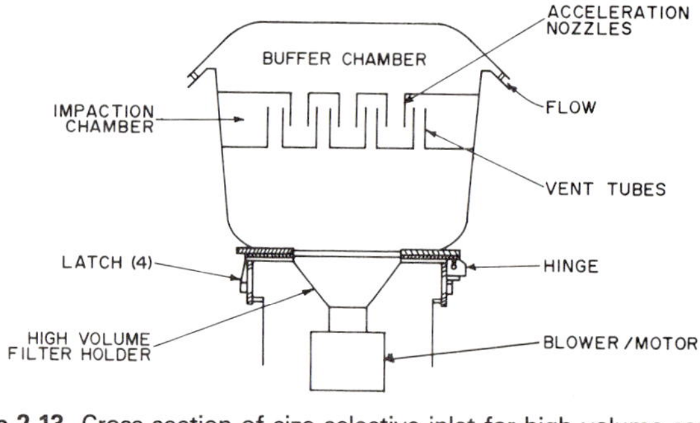

Figure 2.13. Cross section of size selective inlet for high volume sampler.

The other components of the SSI high-volume sampler are identical to those of the TSP high volume and they are described in Section 2.3.4.

2.3.9.2 Operation Procedure ofthe SSI High-Volume Sampler

The operation procedure of the SSI high volume sampler is similar to that of the high volume sampler described in Section 2.3.6. Procedure variations necessary to comply with the construction of the SSI sampler are discussed here.

The first step is to release the spring latches holding the inlet fractionator, tilt it back, and secure it with the cabinet support rod (see Figure 2.14). After loading a new filter element, the inlet fractionator is released from the cabinet support rod, lowered over the filter and secured with the spring latches. The rest of the operation procedures are the same as for the TSP high volume sampler described in detail in Section 2.3.6. The calibration procedures of this sampler are the same as for the TSP high volume sampler.

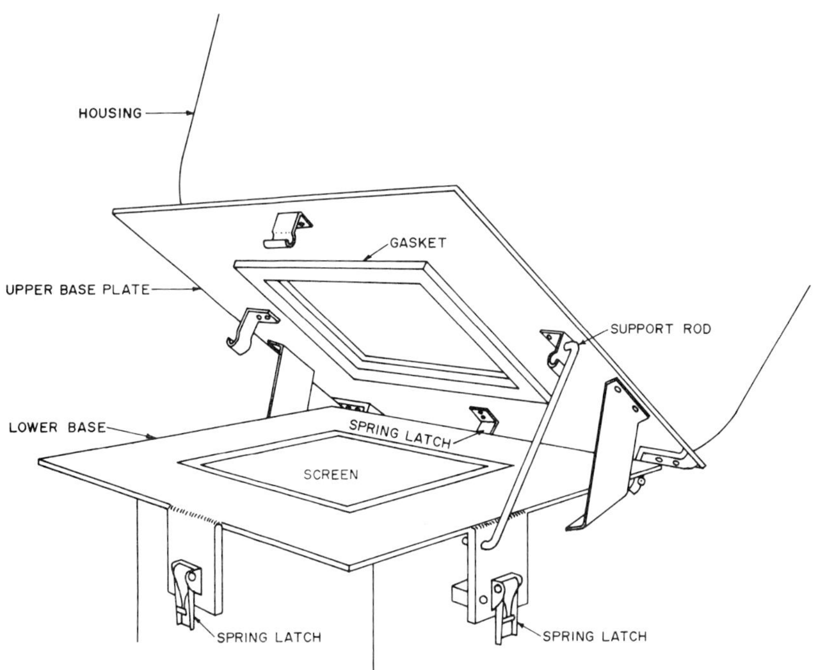

Figure 2.14. Top section of high volume sampler.

2.4/Dichotomous Samplers

2.4.1 Introduction and Precautions

The current Clean Air Act Amendments establish the particulate matter primary standard as 75 $\mu g/m^3$ of total suspended particulates (TSP). Proposed ammendments to the U.S. Federal Register (40 CFR, Parts 50 and 53) relative to ambient air quality sampling also includes a particulate size requirement. For years it was not known whether the inhalable or course fraction would have a 15 or 10 μm upper size limit so many sampler inlets were designed for 15 μm. However, as of mid 1983 it appears that 10 μm will be the size limit and thus the standard has become known as PM-10. This standard specifies performance specifications for the various samplers and manufacturers must supply these test data to the EPA. In addition to TSP, fugitive dust studies may be liable for data on the fine inhalable particulates (IP) of less than 2.5 μm and courses respirable particulates (RP) of 2.5 or 10 (or 15) μm.

To meet the requirements of having instruments capable of fracionating TSP into coarse and fine size ranges, "dichotomous" samplers have been developed. Including the dichotomous samplers, there are three basic types of samplers available for ambient sampling. They operate at (1) low, (2) medium, and (3) high flow rates depending on how much particulate sample mass is desired. These PM-10 samplers by type are:

Sampler	Sample Rate		Particle size Range, μm	Filter Size
	m^3/h	cfm		
Dichotomous	1.0	0.6	0–2.5 and 2.5–10	37 mm
Medium flow	6.8	4.0	0–10	102 mm
Hi-Vol	60	35.3	0–10	25×25 cm

Only the dichotomous samplers give fine particle data. Common samplers of this type are the Sierra Series 240 (244, 244E, and 245) and the Beckman Samplers.

Our discussions normally show 15 μm inlet cut off sizes for these instruments yet the expected PM-10 standard is 10 μm. The sampling instruments involved are basically the same but the inlet heads differ. This is shown in Figure 2.15 where drawing (a) is the older 15 μm inlet and the others are the newer 10 μm heads.

The operator must take certain precautions before using the Dichotomous Samplers to maintain high operating performance and,

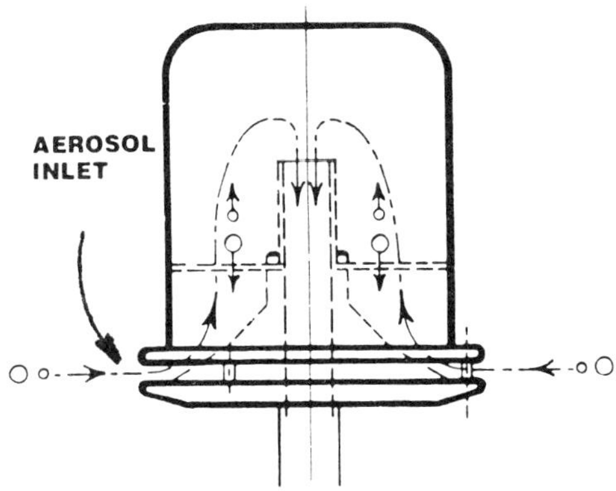

Figure 2.15. Inlet section of dichotomous samplers Model 244 and 244E.

therefore, valid results. These precautions must be taken during preparation of sampler, sampling process, and handling of samples. Detailed specific instructions are given in the following sections, but the following general instructions are listed to assist the operator:

Preparation of Sampler

1. The sampler must be located in open areas with its intake at a height of about 1.5 m above the ground with no obstructions (buildings, trees, etc.) in the immediate vicinity.
2. The sampler's door must be checked for a proper sealing to avoid penetration of the airborne dust into the samplers as the wind blows. Penetrated dust may deposit onto the filter element resulting in serious error.
3. Covering the sampler with polyethylene sheet, leaving only its inlet in the open air would be helpful especially in highly dust loaded locations.

Sampling Process

1. The sampler must turn ON and OFF with ½ hour of midnight.
2. The sampler may operate up to 23 but no more than 25 hours.
3. Decrease in flow rate during sampling of more than 10 percent from the initial value is questionable. Sampling must be terminated and calibration is required.

Handling of Samples

1. After the sampling process the filter holder-carousel (Sierra sampler) and the filter magazine (Beckman sampler) must be removed and carefully placed in specially prepared, sealed compartments to pro-

tect filters from contamination during transportation to the laboratory.
2. The relative humidity in the balance room must be controlled and be less than 50 percent.

2.4.2 The Sierra 244 and 244E Dichotomous Samplers*

2.4.2.1 Description

The sampler divides the air streams passing the 15 μm inlet into two portions that are filtered separately with a total flow rate of 16.7 l/min. The particle size separation is accomplished by pseudo-impaction into an air stream of different velocity and not on an impaction surface. For this reason it is sometimes called a "virtual impactor." The current Sierra dichotomous samplers divide the total sample (0–15 μm) into "fine" (0–2.5 μm) and coarse (2.15–15 μm) fractions, which are collected separately on Teflon filters of 37 mm diameter with a 2–3 micron pore size. The filters are weighed before and after sampling to obtain the mass collected and concentration of each size range. The concentration of the total IP is determined by adding up the concentration of coarse and fine particles.

The Sierra models 244 and 244E are both manual samplers. The model 244 (Figure 2.16) consists of two modules: (1) the sampling module and (2) the flow control module (Figure 2.17). The specification of the Sierra model 244 are given in Table 2.5.

The samplers are equipped with a digital timer/programmer and elapsed time indicator. The digital clock has a LED display and D.C. battery standby. The filter elements are shipped from the lab in nylon cassettes for ease of loading and removal by the operator from the sampler module. The filters remain in these cassettes during sampling and shipment.

The model 244E is identical to the model 244, but it is modified so that it can be calibrated at the control module as well as the sampling module. The model 244E also uses a mechanical timer for operation instead of a digital timer/programmer. This model eventually replaces the model 244 samplers in the IP field investigations.

2.4.2.2 Flow System

The particles are first passed through the inlet section (Figure 2.15) that allows only inhalable particles smaller than 15 μ to enter the virtual

*As of Nov. 18, 1983, Sierra Samplers became part of the Anderson Samplers, Inc. line and are called Sierra-Anderson Samplers.

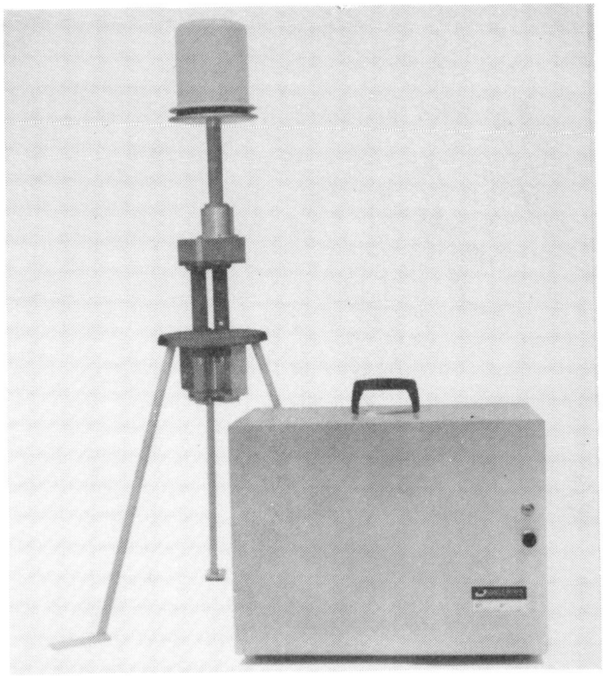

Figure 2.16. Model 244 Sierra dichotomous sampler (Courtesy Sierra Instruments, Inc.).

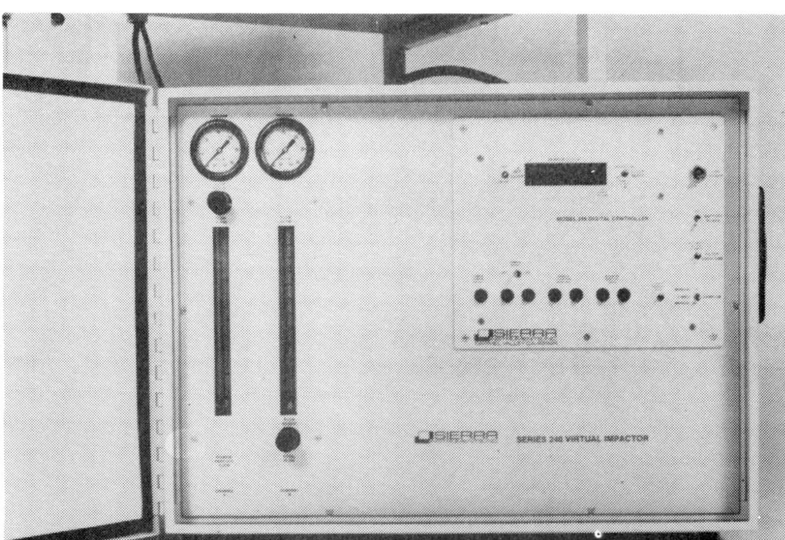

Figure 2.17. Control module of Sierra model 244 dichotomous sampler (Courtesy Sierra Instruments, Inc.).

Table 2.5. Specifications for the Sierra Model 244 Dichotomous Sampler.

Collection efficiency	Mass median diameter at 50 percent collection efficiency for equivalent spherical particles at g/cm³ is 2.5 μm; standard geometric deviation or σ_g = 1.2.
Internal losses	Maximum value over range of 0 to 20 μm is less than 1 to 2 percent and occurs at 2.5 μm. Average loss for all particles is less than 1 percent.
Flow rates	Total Flow: 1 m³/hr, or 16.7 L/min, fine-particle flow: 0.9 m³/hr or 15.0 L/min; coarse particle flow: 0.1 m³/hr, or 1.67 L/min.
Flowmeters	Precision rotameters, ±1.5 accuracy at above flows
Concentration ratio	10:1
Vacuum pump	Diaphragm type, split phase motor, 1/4 hp.
Timer/programmer	Sierra Model 302 Digital Timer/Programmer; built-in; all functions digital and quartz crystal controlled, has digital clock with 1/2-in. LED display and D.C. battery standby; includes first sample period delay up to 9 days, sampler period of 1, 2, 3, 4, 6, 8, 10, 16, 20, or 24 hours, and skipped time between sample periods of 1 to 9 days.
Elapsed time indicator	XXXX.XX hours; nonresettable.
Filter media	37-mm membrane or glass fiber; Teflon membrane media recommended.
Filter holder	Circular, anodized aluminum, 1.750 inches O.D.
Interconnecting tubing	10 m long; 3/8 inch O.D. for fine-particle flow; 1/4 inch O.D. for coarse-particle flow.
Aerosol inlet	15-μm nominal cutpoint over approximately 0 to 20 km/hr wind speed range; includes bug screen.
Power required	115 V AC ±15 percent, 50-60 Hz, 20 W (230 V A.C. ±10 percent, 50-60 Hz, optional-add suffix "x" to model number).
Overall dimensions	Control module: 16″H × 22″W × 11″D; sampling module: 40″H × 25.63″ Dia. tripod base bolt circle.
Net weight	Control module: 50 lb; sampling module: 15 lb.
Total shipping weight	85 lb.

impactor. Figure 2.17 is a photograph of the Sierra 244 control module. The schematic flow system is shown in Figure 2.18, and Figure 2.19 shows the principle of operation of the virtual impactor. The coarse particle flow rate, Q_c = 0.1 m³/hr (1.67 l/min) and is controlled by its flow selector value, which feeds into pressure P_2 at the inlet of the pump. Flow Q_c is relatively constant except for a small decrease in P_4 which can occur during sampling. Because Q_c is small, an error in Q_c can cause an error of only about 10 percent in the particle mass concentration.

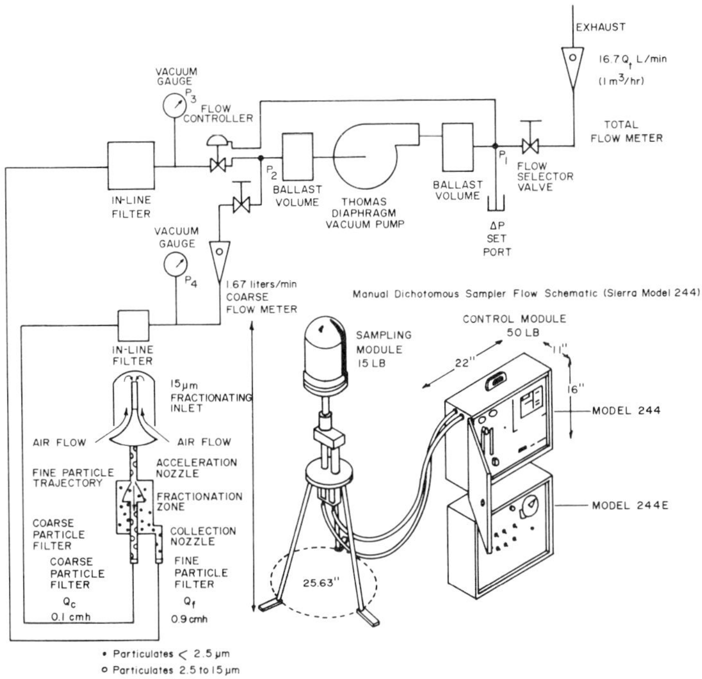

Figure 2.18. Schematic flow system of dichotomous samplers Model 244 and 244E.

2.4.2.3 Control Panel

The control panel is positioned behind a weathertight door and contains vacuum gauge, rotameter, and digital timer/programmer displays. The control module is shown in Figure 2.17. The functions of the various displays and controls are described in Table 2.6.

2.4.2.4 Operation of Sierra Model 244 and 244E Dichotomous Samplers

The operation of the samplers are described in step by step sequence for the general procedures and the timer/programmer in Tables 2.7 and 2.8.

2.4.2.5 Sample Validation

Validation Criteria—In order to assist the operator in determining whether a sample is valid, the following validation criteria have been established for all IP Network samples:

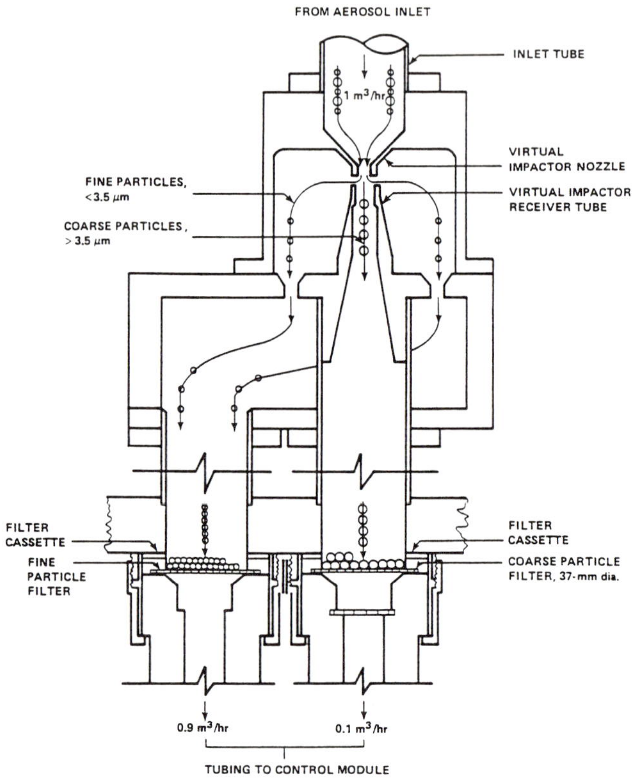

Figure 2.19. Principle of operation of the virtual impactor.

1. Timing—All samplers must turn ON and OFF within ½ hour of midnight. All samplers must operate for at least *23* but no more than *25* hours.
2. Flow Rates—Decreases in flow rate during sampling of more than 10 percent from the initial setpoint are questionable. Changes in flow rate calibration of more than 10 percent, as determined by a field calibration check, will invalidate all samples collected back to the last acceptable flow check.
3. Filter Quality—All particulate deposits that do not have well-defined borders (possible leak) should be voided. Any filter that is obviously damaged (i.e., torn or frayed) should be voided.

2.4.2.6 Operators' Field Calibration Check Procedures

During routine IP Network operation, the operator will be required to check the calibration of the instruments every other sampling period.

Table 2.6. Function of Sierra 244 and 244E, Displays and Controls.

Component	Function
Flow Control System (Flow selector, rotameters, and vacuum gauges	For setting the total and coarse particle flow rates (Q_t and Q_c in Figure 2.17). Vacuum gauges measure the pressure drop across the fine and corase particle filters.
Clock	1. Display LED, hours and minutes; 24-hr format 2. Time Base-Quartz crystal 3. Display switch: (i) Off display is turned off to prolong battery life, all timing functions continue (ii) Time of Day; present time is displayed. (iii) Sample Start Time; Start sampling time is displayed. (iv) Fast/Slow (for setting Time of Day and Sample Start Time); Fast; Minutes advance at 60 Hz rate. Slow; Minutes advance at 2 Hz rate.
Sample After Switch	Sample after X-days. Delays start first sampling period 0 to 8 days (1-day increments)
Sample Every Switch	Permits sampling energy 1 to 9 days (1 day increments)
Sample for Switch	Sets numbers of hours sampler stays on 1, 2, 3, 4, 6, 8, 10, 12, 16, 20, or 24 hours
Sampler Switch	ON: Sampler is turned on manually; timing is unaffected and elapsed timer runs. TIMED: Sampler is controlled by timer. OFF: Sampler is turned off, timer is unaffected.
Set Switch	(Momentary) Normally on "DISPLAY" position "TIMER": Turns sampler off; rests all time functions except present time and "SAMPLE STAR TIME"
Power Fail	Flashing time display digits with "DISPLAY" switch on indicates battery power failure. All timed function must be reset. "AC" POWER FAILURE" light indicates AC line power has failed during a sampling period.
Total Sampling Time Display	Indicates the total elapsed time the sampler is on; non-resettable; 9999.99 hrs.
Power/Circuit Breaker Switch	ON: "Push in;" all power on; timing commences. OFF: "Push up," all AC power off, no sampling, optional continues to run timer.

Calibration checks of the sampler flow rate require the instruments to be running, and hence that timed operation of the master timer be bypassed. Procedures for operation of the master timer and filed calibration checks of the samplers are given below.

Operation of the Tork Time Control (Master Timer)

All samplers are controlled by a master timer to ensure all samplers operate for a 24-hours period every sixth day. The operator does not

Table 2.7. Operation Procedures for the Sierra 244 and 244E Dichotomous Samplers.

Step	Procedure
	A. General Operating Procedure
1	Clean the sampling module and the bug screen.
2	Place the sampling module in the desired place.
3	Load the filters: (i) unscrew the knurled filter holder nuts. (ii) Install the filter cassettes containing the preweighed filters in the sampling module. (iii) Put both filter cassettes on the filter screens. The lower half of the cassette goes over the filter screen. The lower half is also the side having the shortest distance to the filter surface. (iv) Screw on both knurled filter holder nuts tightly (by hand). NOTE: the coarse particular filter is on the center line of the virtual impactor head inlet and the fine particulate filter is offset (see Figure 2.19).
4	Connect the tubes between the Sampling and Control modules and tighten the nuts securely.
5	Place the control module in the desired location (Both modules can be bolted down to their mounted place).
6	Open the front cover of the enclosure of the control module.
7	For Model 244, turn SAMPLER switch on the Digital Timer/Programmer to OFF. For Model 244E, switch mechanical timer to OFF.
8	Plug the cord into line voltage; Model 244: 115/60 Hz 5 Amax Model 244E: 230/50 Hz, 2.5 Amax
9	Be sure the flow selector valve on the bottom of the total rotameter is open.
10	Model 244: Turn SAMPLER switch on the Digital/Timer Programmer to ON.
11	Model 244: depress the power switch to "ON." The pump will turn on.
12	Adjust the rotameter by means of the selector valve (on the total rotameter) to the set point (Figure 2.20) or interpolation Table 2.8 to give a total flow rate of 16.7 l/min. The flow selector will require minor adjustment between samples.
13	Set the coarse flow rate in the same manner using Figure 2.21 and Table 2.9.
14	The sampler (244 or 244E) is now ready for operation. Sampling may be timed by the mechanical master timer or manually.
15	Close the front cover of the control module.
16	After sampling remove the filter cassettes and put them in marked petri dishes.
17	Record sampling time from elapsed time indicator.
18	On Model 244, note if AC POWER FAIL is ON.
	Operating Procedure for Mode 302 Digital Timer/Programmer (Sierra 244)
1	Open cover of enclosure.
2	Turn SAMPLER switch OFF.
3	Plug male cord into line voltage (110 V a.c. for 302, 220 V AC for 302K).

(continued)

Table 2.7. (continued).

Step	Procedure
4	Push POWER switch ON (push in). Pump will not turn on. Set flow controller, etc., of sampler (See other instructions).
5	Set SAMPLE START TIME of Day: (NOTE: Digits will flash until set). NOTE: SAMPLE START TIME must be at least 10 minutes after TIME OF DAY. The DISPLAY Selector Switch must not be in the SAMPLE START TIME position 10 minutes prior to sampling. a. DISPLAY Switch: Set to SAMPLE START TIME. b. FAST/SLOW Switch: Hole up or down as appropriate to set sample start time (24-hour format).
6	Set Present Time of Day (NOTE: Digits will flash until set): a. DISPLAY Switch: Set to TIME OF DAY. b. FAST/SLOW Switch: Hold up or down to set present time of day (24-hour format).
7	Delay Start (Sample: after X Days, 0-8 Days): Set SAMPLE AFTER switch to number of days to be skipped before first sampling period. Position "0" will initiate first sampling period. The first time, the TIME OF DAY = SAMPLE START TIME. Thus, for example, if the present time of day is 10:00 and the start time is 8:00, the first sample will start in 22 hours. Position "1" will delay the start 1 day (24 hours) after TIME OF DAY = SAMPLE START TIME. Position "2" will delay start 2 days (48 hours), etc.
8	Skip Timer (Sample Every Y Days, 1-9 Days): Set the SAMPLE EVERY switch to initiate 1 sample every Y days. Position "1" samples every day. Position "2" samples ever 2 days. Position "6" samples every 6 days, etc.
9	Sampling Period (Sample for Z hours, 1-24 hours): Set the SAMPLE FOR switch for the number of hours the sampler is to remain on each sampling period. NOTE: The switches referred to in steps 9, 10, and 11 are positive detent switches that provide exact timing. If the switch is not in the detent, it is not usable.
10	Set Timer: Push SET switch down to TIMER position for approximate 2 seconds. NOTE: Steps 7, 8, 9, 10, and 11 can be done in any sequence, but Step 12 must be done after Steps 7, 8, 9, 10, and 11.
11	Time Sampling: Place SAMPLER switch in TIMER position. This initializes all timing functions.
	NOTES: (1) Flashing time display indicates a.c. and battery power failure. (2) AC POWER FAIL light (dot) indicates failure of a.c. power during sampling period. (3) Display may be left on; however, standby battery life may be shortened if a.c. power fails. (4) The Digital Elapsed Time Indicator (9999.99 hours maximum, nonresettable) records the total elapsed time the sampler is on (both TIMED and ON positions of SAMPLER switch). A.C. power fail stops elapsed time indicator until power returns. (5) Power switch incorporates circuit breaker. When a circuit is broken, the power ON button pops up. If more than 15 A are drawn, the circuit breaker will trip even if the button is held on (in). Reset is accomplished by pushing ON button in. (6) Manual Operation: The timer can be bypassed by placing the SAMPLER switch in the ON position to independently turn the sampler on during noncycle periods. The OFF position turns both timed and manual sampler power off.

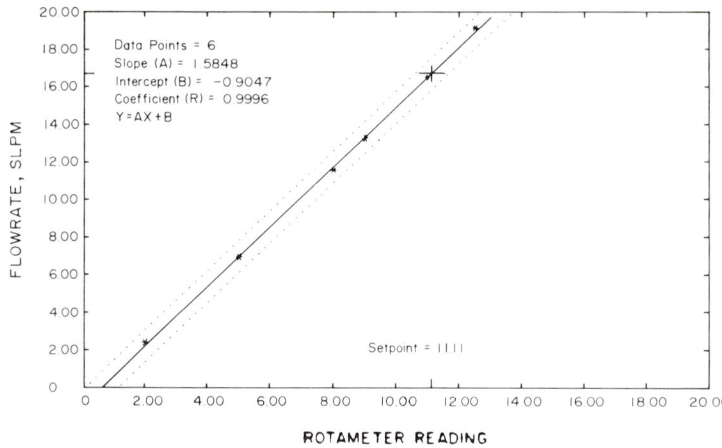

Figure 2.20. Sample total calibration curve for the Sierra dichotomous sampler, inhalable particulate network (Sierra Dichotomous Sampler EPA #174044 S/N 538).

need to be concerned with the master timer except when the timer must be bypassed for field calibration checks, or in the event of a power failure. However, the operator should check the master timer at each sample change to make sure that the next sampling period will be correct.

Bypassing the master timer during field calibration checks. The samplers must be operative during the calibration check. Since the calibration check cannot be accomplished when the equipment is collect-

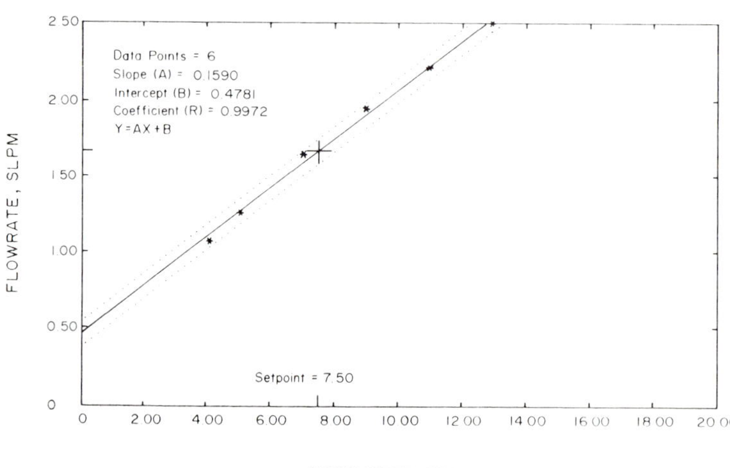

Figure 2.21. Sample coarse rotameter calibration curve for the Sierra dichotomous sampler, Inhalable particulate network (Sierra Dichotomous Sampler EPA #174044 S/N 538).

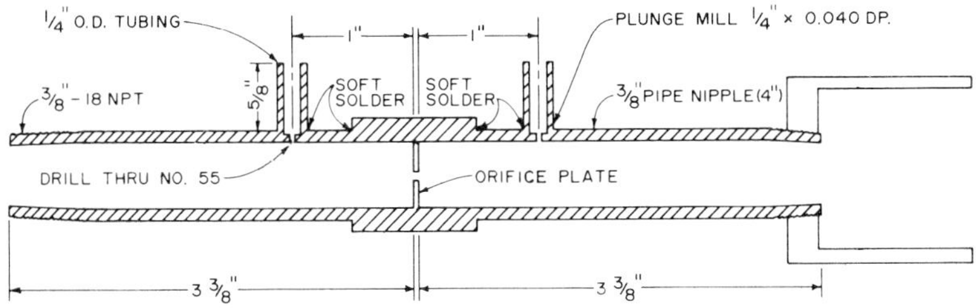

Figure 2.22. Dichotomous sampler calibration check orifice assembly.

ing a sample, the master timer must be bypassed. The timer can be bypassed using the procedures in Table 2.9.

Equipment

The following equipment is required for a field calibration check:

- Calibrated orifice (Figure 2.22)
- Orifice calibration curve (Figure 2.23)
- Sierra dichotomous sampler "coarse" and "total" rotameter calibration interpolation tables (Tables 2.10 and 2.11)
- Magnehelic gauge or manometer

Field Calibration Check Procedure

A field calibration check of the "total" flowmeter is required after every other sampling period for Sierra dichotomous samplers. The check is made by installing an orifice device calibrated in the operating range of the "total" rotameter.

Table 2.8. Seven-Day Mechanical Skip Timer Elapsed Time Indicator (Model 244E).

Step	Procedure
1	Open front door of enclosure.
2	Turn TIMER switch OFF.
3	Plug male power cord of Sampling Module into line voltage (110 V AC, 60 Hz) going to master timer.
4	Set correct time on dial by grasping dial and rotating *clockwise only* until correct day and time-of-day appear at the red pointer.
5	Set the elapsed time indicator to 0000.00 minutes by pressing the rectangular button to the left of the digits.
6	Check the 8-A "slo-blo" fuse by manually turning the sampler on.
7	Leave power on. The master timer will operate the sampler on the scheduled sampling day.

Table 2.9. Procedure for By-passing the Tork Timer.

Step	Procedure
1	Refer to the timer in Figure 2.6.
2	Rotate the skip wheel until the day indicator is pointing to the sampling day (lug removed).
3	Power is now supplied to all samplers.
4	To turn power off, rotate the skip wheel to a no-sampling day (lug in place).
5	When the calibration check is complete, reset the timer as described in the next section.
	Resetting the master timer after power failure or calibration check:
	1. Set the hour dial so that the station time is opposite the hour indicator.
	2. Set the skip wheel so that the number of lugs (clockwise) between the missing lug and the day indicator is equal to the number of days before the next sampling date.

The total flow rate is checked against the field calibration check orifice device. It is expected that in the near future individual checks for the "total" and "coarse" rotameters will be approved and incorporated into this procedure. To perform the field calibration check, follow the procedures given in Table 2.12.

2.4.2.7 Routine Maintenance

Cleaning the Sampling Module:

The Sampling Module is disassembled. All parts are sealed with "O" rings. Particulate internal loss deposits accumulate primarily on the outer and inner surfaces of the tip of the receiver tube in the virtual impactor head. The receiver tube should be inspected periodically for such particulate deposits and cleaned as required. A receiver tube cleaning schedule of every three to four months is typical. The remaining inner surfaces should be cleaned every six to twelve months. Cleaning should be done with alcohol or water using a camel's hair brush or by washing.

The diametral "O" rings in the aerosol inlet and the flow splitting chamber should be conditioned periodically with vacuum grease.

The bug screen in the aerosol inlet should be cleaned periodically during the summer months. The bug screen is exposed for cleaning by pulling the aerosol inlet off the aerosol inlet tube. A diametral "O" ring in the aerosol inlet acts as the seal.

Control Module:

CAUTION: UNPLUG THE LINE POWER CORD FROM ITS RE-

Table 2.10. Sample Interpolation Table for Sierra Coarse Rotameter Calibration.

ROT RDG	SLPM	M^3/Min	ROT RDG	SLPM	M^3/Min	ROT RDG	SLPM	M^3/Min	ROT RDG	SLPM	M^3/Min
4.00	1.1141	0.0011	6.00	1.4321	0.0014	8.00	1.7502	0.0018	10.00	2.0682	0.0021
4.10	1.1300	0.0011	6.10	1.4480	0.0014	8.10	1.7661	0.0018	10.10	2.0841	0.0021
4.20	1.1459	0.0011	6.20	1.4639	0.0015	8.20	1.7820	0.0018	10.20	2.1000	0.0021
4.30	1.1618	0.0012	6.30	1.4798	0.0015	8.30	1.7979	0.0018	10.30	2.1159	0.0021
4.40	1.1777	0.0012	6.40	1.4957	0.0015	8.40	1.8138	0.0018	10.40	2.1318	0.0021
4.50	1.1936	0.0012	6.50	1.5116	0.0015	8.50	1.8297	0.0018	10.50	2.1477	0.0021
4.60	1.2095	0.0012	6.60	1.5275	0.0015	8.60	1.8456	0.0018	10.60	2.1636	0.0022
4.70	1.2254	0.0012	6.70	1.5434	0.0015	8.70	1.8615	0.0019	10.70	2.1795	0.0022
4.80	1.2413	0.0012	6.80	1.5593	0.0016	8.80	1.8774	0.0019	10.80	2.1954	0.0022
4.90	1.2572	0.0013	6.90	1.5752	0.0016	8.90	1.8933	0.0019	10.90	2.2113	0.0022
5.00	1.2731	0.0013	7.00	1.5912	0.0016	9.00	1.9092	0.0019	11.00	2.2272	0.0022
5.10	1.2890	0.0013	7.10	1.6071	0.0016	9.10	1.9251	0.0019	11.10	2.2431	0.0022
5.20	1.3049	0.0013	7.20	1.6230	0.0016	9.20	1.9410	0.0019	11.20	2.2590	0.0023
5.30	1.3208	0.0013	7.30	1.6389	0.0016	9.30	1.9569	0.0020	11.30	2.2749	0.0023
5.40	1.3367	0.0013	7.40	1.6548	0.0017	9.40	1.9728	0.0020	11.40	2.2908	0.0023
5.50	1.3526	0.0014	7.50	1.6707	0.0017	9.50	1.9887	0.0020	11.50	2.3067	0.0023
5.60	1.3685	0.0014	7.60	1.6866	0.0017	9.60	2.0046	0.0020	11.60	2.3226	0.0023
5.70	1.3844	0.0014	7.70	1.7025	0.0017	9.70	2.0205	0.0020	11.70	2.3385	0.0023
5.80	1.4003	0.0014	7.80	1.7184	0.0017	9.80	2.0364	0.0020	11.80	2.3544	0.0024
5.90	1.4162	0.0014	7.90	1.7343	0.0017	9.90	2.0523	0.0021	11.90	2.3703	0.0024

Rotameter setpoint for flowrate of 1.67 = 7.50
L/Min = (0.159014)(rotameter reading) + 0.478055
M^3/Min = (0.000159)(rotameter reading) + 0.000478

Table 2.11. Sample Interpolation Table for Sierra Total Rotameter Calibration.

ROT RDG	SLPM	M^3/Min	ROT RDG	SLPM	M^3/Min	ROT RDG	SLPM	M^3/Min	ROT RDG	SLPM	M^3/Min
8.00	11.7737	0.0118	10.00	14.9434	0.0149	12.00	18.1130	0.0181	14.00	21.2826	0.0213
8.10	11.9322	0.0119	10.10	15.1018	0.0151	12.10	18.2715	0.0183	14.10	21.4411	0.0214
8.20	12.0907	0.0121	10.20	15.2603	0.0153	12.20	18.4299	0.0184	14.20	21.5995	0.0216
8.30	12.2492	0.0122	10.30	15.4188	0.0154	12.30	18.5884	0.0186	14.30	21.7580	0.0218
8.40	12.4077	0.0124	10.40	15.5773	0.0156	12.40	18.7469	0.0187	14.40	21.9165	0.0219
8.50	12.5661	0.0126	10.50	15.7358	0.0157	12.50	18.9054	0.0189	14.50	22.0750	0.0221
8.60	12.7246	0.0127	10.60	15.8942	0.0159	12.60	19.0639	0.0191	14.60	22.2335	0.0222
8.70	12.8831	0.0129	10.70	16.0527	0.0161	12.70	19.2223	0.0192	14.70	22.3920	0.0224
8.80	13.0416	0.0130	10.80	16.2112	0.0162	12.80	19.3808	0.0194	14.80	22.5504	0.0226
8.90	13.2001	0.0132	10.90	16.3697	0.0164	12.90	19.5393	0.0195	14.90	22.7089	0.0227
9.00	13.3585	0.0134	11.00	16.5282	0.0165	13.00	19.6978	0.0197	15.00	22.8674	0.0229
9.10	13.5170	0.0135	11.10	16.6866	0.0167	13.10	19.8563	0.0199	15.10	23.0259	0.0230
9.20	13.6755	0.0137	11.20	16.8451	0.0168	13.20	20.0147	0.0200	15.20	23.1844	0.0232
9.30	13.8340	0.0138	11.30	17.0036	0.0170	13.30	20.1732	0.0202	15.30	23.3428	0.0233
9.40	13.9925	0.0140	11.40	17.1621	0.0172	13.40	20.3317	0.0203	15.40	23.5013	0.0235
9.50	14.1509	0.0142	11.50	17.3206	0.0173	13.50	20.4902	0.0205	15.50	23.6598	0.0237
9.60	14.3094	0.0143	11.60	17.4790	0.0175	13.60	20.6487	0.0206	15.60	23.8183	0.0238
9.70	14.4679	0.0145	11.70	17.6375	0.0176	13.70	20.8071	0.0208	15.70	23.9768	0.0240
9.80	14.6264	0.0146	11.80	17.7960	0.0178	13.80	20.9656	0.0210	15.80	24.1352	0.0241
9.90	14.7849	0.0148	11.90	17.9545	0.0180	13.90	21.1241	0.0211	15.90	24.2937	0.0243

Rotameter setpoint for flowrate of 16.70 = 11.11
L/Min = (1.584809)(rotameter reading) - 0.904737
M^3/Min = (0.001585)(rotameter reading) - 0.000905

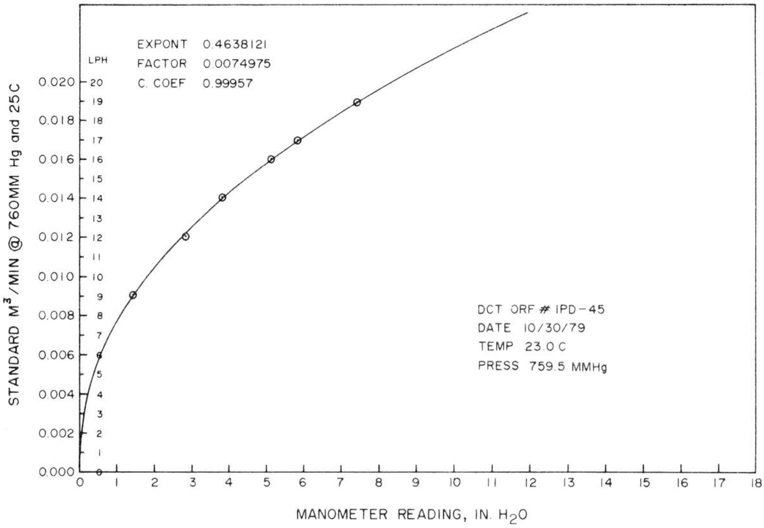

Figure 2.23. Sample dichotomous flow orifice calibration curve from inhalable particulate network.

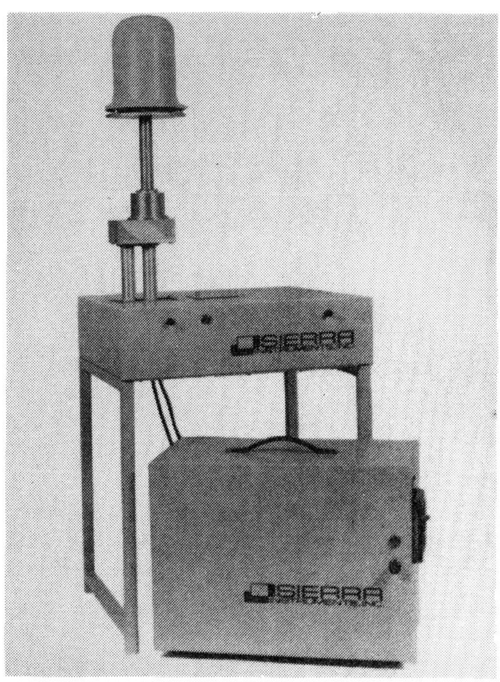

Figure 2.24. Model 245 Sierra dichotomous sampler (Courtesy Sierra Instruments, Inc.).

Table 2.12. Field Calibration Check for Sierra 244 and 244E.

Step	Procedure
1	Insert clean filters into both the "fine" and "coarse" filter cassettes in the sampler as described in the operation procedure (Section 2.4.2.4).
2	Remove the standpipe from the sampler and replace with the "total" flow calibration check orifice device.
3	Turn on the sampler and allow it to warm up to operating temperature (approximately 5 minutes).
4	Open both "total" and "coarse" flow control valves full counterclockwise. Adjust both the "total" and "coarse" rotameters to their respective setpoints as recorded on the laboratory calibration curves or interpolation tables (2.10 and 2.11) provided with the sampler.
5	Observe the pressure drop, ΔP, across the orifice, and its corresponding flow rate from the calibration data (Figure 2.23) provided with the orifice. Record both values. Also record the rotameter setpoints and their corresponding flow rates on the Flow Check Data Sheet.
6	Using the above information and the formulas provided in the Flow Check Data Sheet, calculate the Q_c Check %. Record this value on a Flow Check Data Sheet.
7	If the calculated Q_c Check percent is within ± 10 percent of the 16.7 L/min total flow rate, the sampler is operating properly.
8	Turn off the sampler, remove the orifice device, and replace the standpipe.
9	Remove the filters from both "fine" and "coarse" cassettes.
10	Set up the sampler for the next sampling period according to the operating procedure in Section 2.4.2.4.
11	A calculated Q_c check percent not within ± 10 percent of the 16.7 L/min total flow rate usually indicates that the fine flow filter is not sealed properly. Using gloves, gently push back and forth on the filters. If one is not properly sealed, it should snap into place. The pressure drop (ΔP) across the orifice device should now yield a Q_c Check value within 10 percent of the setpoint, 16.7 L/min. If the Q_c Check value is still outside the ± 10 percent range, try another set of filters. If this does not yield a Q_c Check value within the interval 16.7 L/min ± 10 percent, the sampler requires recalibration. Record the value on the Flow Check Data Sheet.

CEPTACLE BEFORE REMOVING THE FRONT PANEL OF THE MODEL 244 CONTROL MODULE ENCLOSURE.

Filter elements: After approximately 12 to 24 months of sampling, the inlet filters for the coarse and fine particle flows may need replacement. In normal operation, air entering these filters is filtered by the two 17-mm membrane filters in the Sampling Module. Hence, inlet filter replacement is infrequent. To replace these filters, unplug the line power cord and remove the front panel of the Control Module by removing the six screws. The filter jar for the coarse-particle flow is the small jar (approximately 1.5 in. dia. × 3 in. L) in the upper left side of the enclosure.

Unscrew the jar, remove the old filter element, and replace with a new element. Tighten the jar very tightly when reinstating to avoid leaks. The fine-particle filter jar is behind the coarse-particulate filter jar. It is approximately 3 in. dia. × 5 in. L and is the one closest to the bulkhead fitting on the side of the enclosure. Filter element replacements are available from Sierra.

Vacuum pump: The diaphragm of the Model 727CA418 Diaphragm Vacuum Pump is replaced routinely at 1- to 2-year intervals or if sudden reductions in sampler vacuum occur and a leak check indicates there are no leaks in the system.

To replace the diaphragm, unplug the line voltage and remove the front panel. Remove the finned head of the pump by removing the six head screws. Remove the four diaphragm plate holddown screws and change the diaphragm. To reassemble, reverse the procedure making sure that the screw clearance cavity in the plate is lined up under the intake valve screw heads and that all head screws are tightened evenly.

2.4.3 The Sierra 245 Dichotomous Sampler

2.4.3.1 Description

The Sierra 245 dichotomous sampler works on the same principles as those of model 244 and 244E (Section 2.4.2). The latter are manually operated and the Sierra 245 is an automatic sampler which is capable of changing the filters after sampling. Figure 2.24 is a photograph of the 245 sampler. The automatic filter changer allows unattended operation of the sampler for weeks or months. The filter holder (carousel) accommodates 20 filter pairs of the same size as those used with models 244 and 244E and they are automatically indexed into the rotary filter holder carousel. The number of filter pairs is selectable from 0 to 20 for which the sampling period can be from 1 to 99 hours. The (skip) time between sample can be adjusted to be from 1 to 99 hours or from 0 to 99 days. A time delay unit is provided so that the start sampling sequence is selectable from 0 to 9 days. A digital display in the control module shows the present time of day as well as the sampling start time. The sampler is equipped with a digital controller that automatically indexes a fresh filter pair into the sampling head if the filters are overloaded and it runs for the remainder of the preselected sampling. A flow event recorder is attached to the side of the control module which gives an independent hard copy record that the sampler turned on and did sample as programmed. The flow is controlled in this model to be constant to 35 cm Hg rather than 20 cm Hg. In the event of AC power interrupt, the standby battery provided with the instrument maintains the time keeping function up to 10 days. As for models 244 and 244E, the sampler consists of two main parts, sampling module which contains the sampling inlet head, virtual impactor, and the filter holder carousel. The second part is the control module which contains the control devices, digital display, flow meters, and the flow event recorder. The cable and PVC tube which are connected between the modules are about 10 m in length, so that the sampling module can be placed in the sampling location while the control module is kept in a closed room for protection against the environment. The general specifications of Sierra 245 Dichotomous sampler is given in Table 2.13.

2.4.3.2 Sampling Module

The automatic 245 Dichotomous sampler consists of the sampling and control modules. The sampling module consists of the particles inlet, virtual impactor head, two filter holder cassettes in engagement with the virtual impactor head, rotating filter cassette carousel, filter plunger mechanism, heater, enclosure, and four mounting legs. The particle inlet is bell shaped, rain-proof, and omnidirectional. It has a built-in bug screen and has a 15 μm cut point particle inlet 1.2 m from the ground. The particles are drawn into the sampler through the inlet at a flow rate of 1 m^3/h (16.7 l/m). The virtual impactor is identical to that used with model 244 and 244E and both coarse (2.5 to 15 μm) and fine (<2.5 μm) particles are collected on separate filters. The coarse particles are collected on the filter at a flow rate of 0.1 m^3/h (1.67 l/min) while the fine particles are collected at a flow rate of 0.9 m^3/h (15 l/min). The uniformity of particulate deposit on both filters is about 5 percent. The filters are enclosed in polypropylene filter cassettes for protection during removal, transportation, and storage. Teflon filters with 2 or 3 μm pore size are the most suitable types to be used with this sampler, however, other kinds such as membrane or cellulose-fiber can be used.

The sampling module has a welded, painted aluminum enclosure that contains the filter cassette carousel, rotating mechanism, and filter plunger mechanism. The circular carousel is a PVT removal plate of a diameter 50.8 cm with 20 holes for accommodating up to 20 pair of filter cassettes. After loading the carousel with fresh, clean, pre-weighed filters, it is inserted into the sampling module, and removed after sampling by opening the lockable door at the top and sliding it on the two guide rails.

Index marks on center of the carousel indicate the number of the filter pairs being exposed. The carousel can be advanced automatically to position "1" by means of a switch in the enclosure. This is usually done when positioning the carousel for start of sampling by filter pair number

Table 2.13. Specifications of the Sierra Model 245 Dichotomous Sampler.

Collection Efficiency	Mass median diameter at 50 percent collection efficiency for equivalent spherical particles at 1 gm/cc is 2.5 microns.
Aerosol Inlet	15 micron nominal cut-point over approximately 0–20 KPH wind speed range; includes bug screen.
Internal Losses	Maximum value over range of 0–20 microns is less than 5 percent and occurs at 2.5 microns.
Flow Rates	Total flow: 1 cubic meter per hour (1CMH), or 16.7 LPM; fine-particle flow, 0.9 CMH, or 15.0 LPM; coarse-particle flow: 0.1 CMH, or 1.67 LPM.
Filter-Holder Carousel	Holds 20 filter pairs; 50.8 cm (20 in.) Dia. × 0.9 cm (0.38 in.) Thick; material-PVC; weight 2.3 kg (5 lb).
Flow Meters	Precision rotameters, calibrated to ±3 percent accuracy at above flows.
Accuracy of Flow Control	±5 percent over a pressure drop range of 0 to -35 cm Hg. across fine-particle filter.
Vacuum Pump	Diaphragm type, split phase motor, ½ h.p.
Digital Controller	All functions digital and quartz-crystal controlled; has digital clock with ½″ LED display; timing accuracy: ±2 min/week; see Description for functions.
Elapsed Time Indicator	XXXX.XX hours; non-resettable.
Filter Media	37 mm membrane or glass fiber; Teflon membrane media recommended.
Filter Holder	Circular; Lexan; 2″ × 2″ filter slides optional
Environmental	Meets EPA requirements; temperatures: -20 to +40C; barometric pressure: 600–700 mm Hg.
Power Required	115 VAC ±10 percent, 50–60 Hz, 600 watts max.: optional: 230 VAC ±10 percent, 50 Hz, add suffix "X" to model number.
Overall Dimensions	Control Module: 16″ H × 22″ W × 11″ D; Sampling Module: 56″ H, 23.50″W × 21″ D; aerosol inlet height 1.2 m; four ¼-20 mounting bolts on the corners of a 22.50″ W × 16.25″ D rectangle.
Net Weight	Control Module: 25 Kg (55 lb); Sampling Module: 18.5 Kg (41.5 lb).
Total Shipping Weight	55 Kg (120 lb).

1. As the sampling starts, a synchronous AC motor turns a cam which raises the filter plunger with the two filter cassettes bringing them in tight contact with the virtual impactor head.

The sampling process is performed automatically according to the following controlled sequences:

1. About 15 seconds before the exact programmed time of sampling

start, the filter plunger rises and engages the filter cassettes, sealing against the virtual impactor head.
2. The pump turns on and the sampling period begins.
3. At the end of the programmed sampling period, the pump turns off.
4. The filter plungers lower to a position below the carousel.
5. The carousel advances to locate the next pair of clean filters in position ready for the next sample.

This cycle is repeated until the 20 pair of filters are all exposed.

2.4.3.3 The Control Module

The control module contains two rotameter flow meters for coarse-particle flow and the total flow, flow selector values for adjusting the flow rates, vacuum gauge above each flow meter and digital programmer. The digital programmer (Figure 2.25) has a modular plug for all exterior connections and it can easily be removed with its panel from the control module.

The flow event recorder is fixed on the side wall of the control module. The enclosure includes also ½ h.p. diaphragm vacuum pump, constant flow controller, and cooling fan. The accuracy of the total flow meter is about 2 percent as the meter reads a 16.7 l/min while the total flow range is of 0 to 20 l/min. The coarse particle flow meter has the same accuracy as it reads 1.67 l/min and its total range is 0 to 2 l/min. The flow meter reading should be corrected for altitude if necessary.

2.4.3.4 Flow Control System

Figure 2.26 shows the schematic of the flow system in the Sierra 245 Sampler. It can be noted that as the filters collect particles, pressure drop (P_1–Pa in Figure 2.26) across them takes place and increases more rapidly with accumulation of particles. The flow control is essential in this sampler so that the design conditions for particle fractionation and collection are maintained. Therefore, the sampler is provided with a flow controller which throttles the flow on the inlet (vacuum) side of the pump and is capable of maintaining a constant pressure drop P_1–Pa by means of a pneumatic feedback line. This ensures that the flow rate is maintained within 5 percent of the collector design characteristics. Exceptions are for the cases in which the atmospheric air density varies during the sampling period. Other variations in atmospheric sampling occur when air temperatures change during sampling. However, a constant temperature controller (Figure 2.26) can be provided to keep the total flow at a constant temperature of 25 $\pm$ 5 °C.

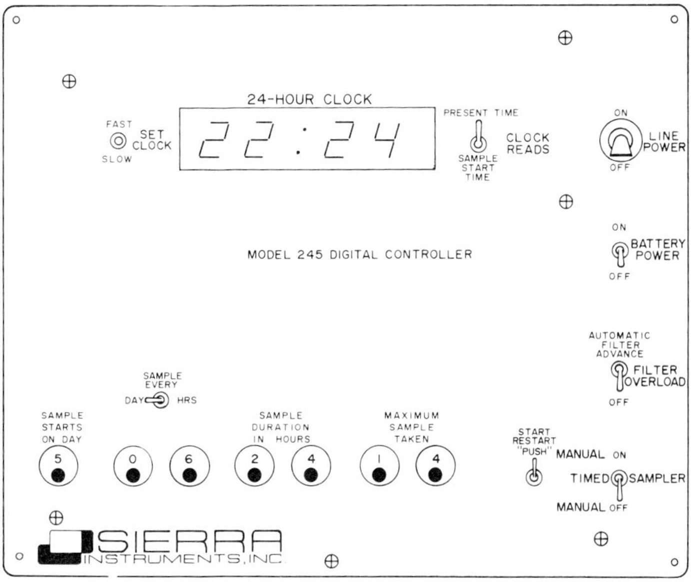

Figure 2.25. Front panel of digital programmer.

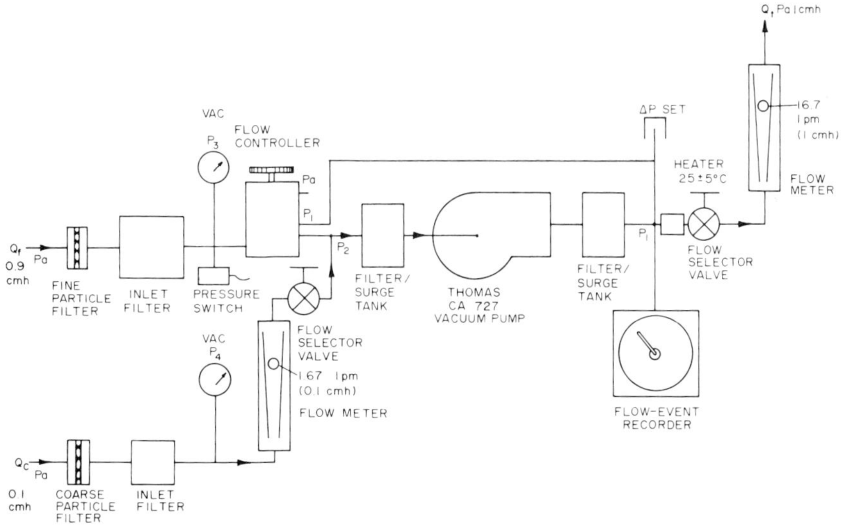

Figure 2.26. Schematic of Sierra 245 flow system.

If the pressure drop across the fine filter ($Pa-P_3$) increases beyond 28 ± 2.5 cm Hg, the limit to which the flow can be maintained constant, the pressure switch shown in Figure 2.26 activates and the digital programmer automatically indexes a clean filter pair into the sampling position. The remainder of the preselected sampling period is then completed with the new filters, and this will be indicated on the flow event recorder.

2.4.3.5 *Flow Event Recorder*

This device is a pressure transducer recorder with a pressure range of 0 to 6 in. (15.2 cm) Hg. It is capable of recording the pressure drop P_1-Pa across the flow selector value on the exhaust side of the pump. The flow is controlled constant at a nominal value of 4 ± 1 in. (10.2 ± 2.5 cm) Hg. The recorder will read 4 in Hg whenever the sampling pump is on and will be zero when it is off. The recorder will also show a zero value when the pressure switch senses filter overload. It is, therefore, possible for the operator to know the time at which the filter overload took place. However, the exact time is not necessary in the subsequent analysis as long as the extra filters have been used for the remainder of the preselected sampling period.

2.4.3.6 *Preparation Procedure Before Sampling*

There are some necessary procedures that must be followed before selecting the sampling program and setting the sampler into operation. First, the installation and setting up the sampler; second, the filters have to be prepared, weighed, and loaded; third, setting the flow event recorder, and finally adjusting the flow rates (the total and the coarse particle flow rates). Upon completion of these main preparation steps, the sampling program can be selected and the sampler is ready for operation.

The details of the preparation steps are given in Table 2.14, the preparation and loading of filters in Table 2.15, setting the flow event recorder in Table 2.16, and adjusting the flow rates in Table 2.17.

2.4.3.7 *Sampling Program*

The detailed sampling program for the Sierra 245 is listed in a stepwise fashion in Table 2.18.

Table 2.14. Procedure for Setting up the Sierra 245 Sampler.

Step	Procedure
1	Assemble legs to bottom of sampling module using the eight ¼-20 bolts and washers provided.
2	Assemble impactor body to sampling module being sure to capture the Parker No. 2-156 "O" ring between the top plate and bottom of the impactor body. Fasten with (4) 10-32 screws provided.
3	Place the bell shaped inlet section on inlet tube and press down until seated.
4	Connect polyflow tubing to their respective inlet/outlet ports. The coarse-particle flow (1.67 l/min) goes through the 0.25 in O.D. tubing and the fine-particle flow goes through the 0.375 in O.D. tubing (the difference in sizes prevent any possible mix-up). Tighten finger tight, then ¾ turn with wrench.
5	Connect twin cable to control module. Connector is keyed and will only fit one way.

Table 2.15. Preparation and Loading of Sierra 245 Filters.*

Step	Procedure
1	Separate the two halves of the polypropylene filter holder (Model FH-240-P) using the thumb nail.
2	Weigh the filter element using a micro-balance to the nearest microgram.
3	The filter is carefully placed into the pocket of the lower female half of the holder. If teflon filters are to be used, then make sure that the smooth teflon membrane side is up and the rougher (non-woven polypropylene) side is down.
4	Place the top male half of the holder over the filter and press the two halves together tightly by hand until they snap together.
5	Number the filter holder for identification with a marker pen. Don't use any stickers over the holder for marking as they can cause leaks.
6	Place the filter holder carousel on a table with the pockets for the filter holders up and place the loaded filter holder into the pocket with the female half of the filter holder down.
7	Make sure that the filter numbers coincide with the index marks on the carousel and that the filter holders for fine particulates are in the outside ring and that for coarse particulates are in the inside ring. The polypropylene filter holders are usually provided in two colors (white and yellow). One color can be used for fine particulates and the other for the coarse particulates. If all holders having the same color, a numbering system of 1F to 20F and 1C to 20C for the carousel can be used.

*Note: The filters must be loaded after being weighed in the balance room which must be highly clean.

Table 2.16. Setting the Sierra 245 Flow Event Recorder.

Step	Procedure
1	Make sure that the recorder pen works satisfactory, that is by lifting the pen, removing the cap from its tip, placing a piece of scrap paper under rim and move the paper against tip to initiate ink flow. If, however, the pen is dry, moisten the tip with water, or it may need replacement with a fresh one if the ink runs out.
2	Install a new chart by lifting the pen using pen lifter, slip chart under raised pen arm and locate chart hole over hub, slip edge of chart under the chart guide and time index clip and then push the chart flat against dial plate.
3	Using a screwdriver to rotate chart hub until the correct time scale line on the chart coinsides with the time index clip (pointed clip).
4	Adjust zero using "zero adjust screw."

2.4.3.8 Data Analysis

Care must be exercised in the handling, preparing, and weighing of the sample as described in Table 2.19. Then correct the flow rates for the local altitude of the sampling location by multiplying the flow indicated on the rotameters by the correction factors which can be determined by

Table 2.17. Adjusting Sierra 245 Flow Rates.*

Step	Procedure
1	Open the door of the sampling module enclosure.
2	Insert loaded filter holder carousel into the front of the sampling module keeping it level. Be certain the carousel is on top of the rails and well inside the enclosure.
3	Center the carousel over the large central pen of the motor drive shaft and rotate it so that the small hole in the carousel is aligned with the small pin on the motor drive shaft. Then insert the carousel coupling into hole in carousel, lining up small hole with pin so that it seats all the way. A small back and forth motion may be necessary to seat coupling fully, then close and lock the door of the sampling module.
4	With all four right hand switches off (down position) plug unit into line voltage.
5	Turn "LINE POWER" switch on, push "SET CLOCK" momentarily to stop clock flashing and push "START/RESTART." Move "SAMPLE" switch to "MANUAL ON." Plungers should now move upwards to engage filter pair and pump should start.
6	Set total flow rate. Turn the flow adjust valve marked "TOTAL FLOW" and select flow rate of 16.7 l/min. The vacuum gauge will read between 5 in to 20 in Hg, depending on filter media used.
7	Set coarse particle flow rate. Turn the flow adjust valve marked "COARSE PARTICLES FLOW" and select flow rate of 1.67 l/min. The vacuum gauge will read approximately zero.

*Note: For adjusting the flow rates, the filters must be in place.

Table 2.18. Sampling Program for Sierra 245 Instrument.

Step	Procedure
1	Move all right hand side panel switches to off position.
2	Plug line cord into line voltage.
3	Turn "LINE POWER" and "BATTERY POWER" switches "ON."
4	Set clock to present time by moving "CLOCK READS" switch to "PRESENT TIME" position and advance clock with "SET CLOCK" switch to present time. Moving this switch up advances clock rapidly for hours adjustment and down advances clock slowly for minutes.
5	Set sample start time by moving "CLOCK READS" switch to "SAMPLE START TIME," and set clock as in (4). NOTE: Sample start time must be at last 20 minutes ahead of present time or it will be ignored for 24 hours. Return "CLOCK READS" switch to "PRESENT TIME" position after setting sampler start time, or machine will not operate.
6	For sampling to start today, set "SAMPLE STARTS ON DAY" rotary switch to "1." To delay timed sampling for up to 8 days, set "2" for tomorrow, "3" for the day after tomorrow and so on up to "9." DON'T USE ZERO (invalid).
7	Set "SAMPLE EVERY" switch on "DAY" and rotary switches to "01" for every day, to "02" for every two days and so on.
8	Set "SAMPLER DURATION" to "01" for one hour, "02" for two hours, and so on.
9	Set "MAXIMUM SAMPLES TAKEN" to the number desired.
10	Momentarily depress "START-RESTART" to initialize program.
11	Move "SAMPLER" switch to "TIMED."
12	Set "FILTER OVERLOAD" switch to "AUTOMATIC FILTER ADVANCE," if desired.
13	Load filter holder carousel.
14	Move "HOME" switch located inside front door of sampling module to "HOME" position and hold until carousel stops at number 1 position. Home switch is a momentary type to prevent it from being left in "HOME" position.
15	Close and lock the front door of the control module.
16	The unit will now automatically change the filters as programmed and is ready to turn.

means of Figures 2.27 and 2.28 for the total Q_t and the coarse particles Q_c flow rates respectively (the fine particle flow rate, $Q_f = Q_t - Q_c$). An example using Figures 2.27 and 2.28 at 1,000 ft altitude for initial settings is:

$$Q_c = \frac{1.6}{0.824} = 2.03 \ \ell\text{pm}$$

Table 2.19. Preparation of Sample and Data from Sierra 245.

Step	Location	Procedure
1	End of Sample	Carefully take the filter holder carousel outside of the sampling module and transport to the Laboratory in an enclosed compartment.
2	Balance Room	Remove the filter holders from the carousel's pockets and desiccated and allow the filters to equilibrate with the room air for about 24 hours.
3	Filter Holders	Split the two halves of the filter holder and weigh filters one by one using the microbalance to the nearest microgram and record the weights.
4	Calculation	Determine the weight of particulate material collected on each filter by subtracting the filter weight before and after sampling.

$$Q_f = \frac{16.7}{0.809} = 20.6\ \ell pm$$

Calculate the average fine and coarse particulate concentrations over the sampling time, t as:

$$C_f = \frac{M_f}{Q_f t}$$

$$C_c = \frac{M_c - C_f Q_c t}{Q_t t} = \frac{1}{Q_t t}\left(M - \frac{Q_c}{Q_f} M_f\right)$$

Therefore, the total inhalable particulate concentration C_I is:

$$C_I = C_f + C_c + \frac{M_f + M_c}{Q_t t}$$

The above equation shows that an error of ± 10 percent in Q_c will cause an error of only ± 1 percent in Q_c. Thus, constant flow control of Q_c is not as critical as Q_t.

Chemical analysis of filters may be carried out after the determination of particulate concentration, such as X-ray Flourescence or Atomic Absorption Analysis.

2.4.3.9 Calibration Check of the Sampler

1. Leak Check

The operator will be required to check the calibration of the sampler which is mainly a flow calibration (total, fine-particle, and coarse parti-

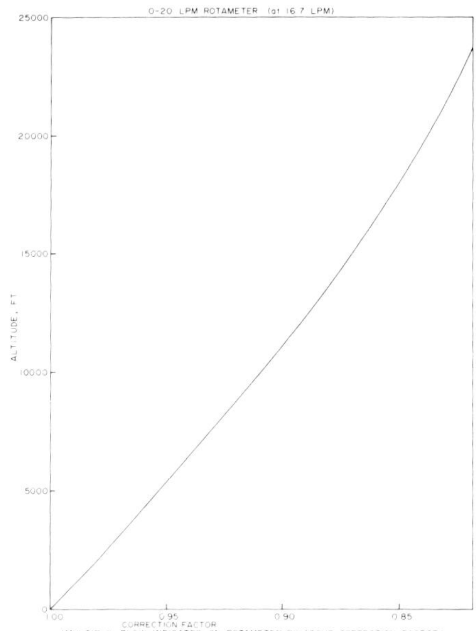

Figure 2.27. Altitude correction factor for series 240 samplers, 0-20 LPM rotameter, at 16.7 LPM (Courtesy Sierra Instruments, Inc.).

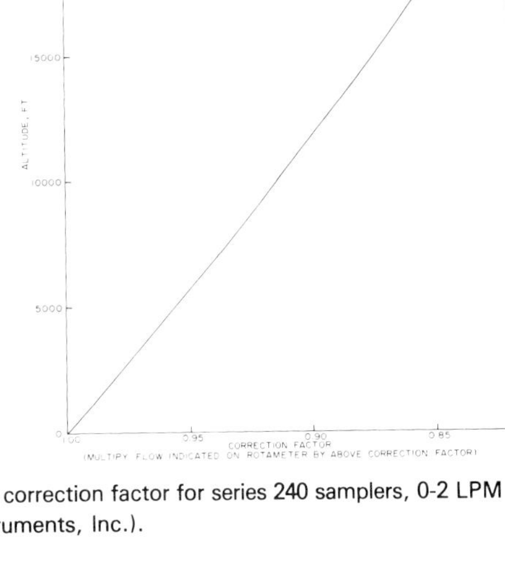

Figure 2.28. Altitude correction factor for series 240 samplers, 0-2 LPM rotameter, at 1.67 LPM (Courtesy Sierra Instruments, Inc.).

cle flow). The calibration may be checked every other sampling period. Every calibration check of the instrument must be preceded by a leak check to assure that leaking does not affect calibration. The procedures necessary to leak check the main component of the sampler and to calibrate it are given in Table 2.20.

Sierra Model 245 measures the coarse particulate flow rate, Q_c, and the total flow rate, Q_T. The fine particulate flow rate, Q_f, is the difference, $Q_T - Q_c = Q_f$. These flows are measured with precision glass bore rotameters. The Q_T rotameter has a scale of 0 to 20 l/min and is calibrated with air at STP conditions (25 °C, 760 mm Hg) with a special calibration point at 16.7 l/min.

The Q_c rotameter has a scale of 0 to 2.3 l/min and is calibrated with air at STP conditions (25 °C, 760 mm Hg) with a special calibration at 1.67 LPM. The Q_c and Q_T rotameters are both calibrated by an independent lab.

Each rotameter is provided with a correction factor for altitudes from 0 to 20,000 feet above sea level (See Figures 2.32 and 2.33). These curves encompass the barometric pressure range of 600 to 760 mm Hg. Using the curves is a simple matter as shown in the following example:

Example: Assume the Model 245 is located at an altitude of 5,000 ft. The two rotameters on the front panel should be set to the following indicated values on the rotameters:

"Coarse Flow": For coarse particle flow, using 1.67 as desired standard flow and looking across the altitude correct factor curve (Figure 2.33) at 5,000 ft., we see a value of .912.

$$\frac{1.67}{.912} = 1.83\ \ell\text{pm}$$

This indicates you should adjust the flow to read 1.83 l/min for an equivalent standard flow of 1.67 ℓpm.

"Total Flow": The same procedure is used for the total flow which yields (use Figure 2.32):

$$\frac{1.67}{.910} = 18.4\ \ell\text{pm}$$

Q_T and Q_c are measured directly in the Model 245 instead of Q_f because this method is more accurate than measuring Q_c and Q_t. The total flow rate Q_T can be calibrated in two locations—at the inlet of the system and the exit of the system. Flow Calibration at the inlet is best because it checks the entire system. This is described in Table 2.20.

Table 2.20. Procedures for Leak Check and Calibration Sierra 245 Sampler.

Step	Procedure
	I. Leak Check
	A. Control Module: NOTE: Leak check should be conducted at the module before calibration to assure that leakage does not affect calibration.
1	Plug both the fine-particle flow and coarse-particle flow bulkhead female pipe threads with either pipe plugs or with your fingers.
2	Turn unit on manually and allow vacuum gauges to reach 20 to 25 in. in Hg.
3	Turn off pump.
4	Pointers on the vacuum gauges should drop from 20 in. Hg to 5 in. Hg in approximately 10 to 12 seconds. If the rate of vacuum loss is greater than this, check for leaks using the following procedure: NOTE: If pressure does not hold required time, unplug Control Module from electrical outlet and follow this procedure: 1. Remove instrument panel of Control Module. 2. Check all tubing fittings inside Control Module for tightness. 3. Check all filter jars for tightness. 4. Replace instrument panel. 5. Plug cord into electrical outlet. 6. Repeat above procedure. NOTE: If pressure drop holds, go to reassembly and leak-checking of entire system. If pressure drop does not hold for required time, call Sierra Instruments representative.
	B. Reassembly and leak-checking entire system:
1	Install the tubing fittings and tubing as they were.
2	Remove Aerosol Inlet from inlet tube and install Model 247 Total Flow Audit Adapter over the inlet tube making sure the "O" ring is in engagement.
3	Turn pump on manually.
4	Plug tube fitting of Total Flow Audit Adapter with your finger.
5	Both vacuum gauges should reach 20-25 in. Hg vacuum.
6	Turn off pump and observe rate of vacuum drop.
7	Pointer should drop from 20 in. Hg to 5 in. Hg in approximately 10 to 12 seconds. NOTE: If rate of vacuum loss is greater, check for leaks using the following procedure: Disconnect tubing from enclosure of Sampling Module and plug ends of tubing with your fingers. If rate of vacuum loss is acceptable, the problem is in the impactor head and not in the tubing.
	C. Impactor Leak Check:
1	Make sure filter holders, seals, and "O" rings are in place and not cut or marred.
2	Tighten all loose screws.
	D. Tubing/Fitting Leak Check:

(continued)

Table 2.20. (continued).

Step	Procedure
1	Visually inspect tubing for cracks or holes.
2	Visually inspect tubing fitting ferrules for cracks.
	NOTE: The correct way to tighten the polyflow fittings is to tighten ¾ of a turn past finger tight.
	II. Flow Calibration
	A. Calibration of Inlet:
1	Remove the Aerosol Inlet from the 1.25 in. O.D. inlet tube by pulling it off vertically. An "O" ring in the Aerosol Inlet effects the seal.
2	Place the Model 247 Total Flow Audit Adapter (to be provided) or any other suitable adapter over the inlet tube and push down vertically. An "O" ring in the Model 247 effects the seal.
3	Connect the hose bib on the Model 247 with tubing to the calibration flow meter.
4	Turn pump on. Turn "FLOW ADJUST" knob above "TOTAL FLOW" on the instrument panel of the Control Module to set the precise 1.67 l/min flow rate.
5	Record the reading on the "TOTAL FLOW" rotameter. This is an accurate "transition standard" for the calibrated flow rate.
	B. Calibration at Exit:
1	Remove the instrument panel of the Control Module.
2	Locate the total flow exhaust port on top of the total flow rotameter.
3	Screw a tube fitting into this port and connect with tubing to the calibration flowmeter.
4	Turn pump on. Turn "FLOW ADJUST" knob above "TOTAL FLOW" on the instrument panel of the Control Module to set the precise 16.7 l/min flow rate.
5	Record the reading on the "TOTAL FLOW" rotameter. This is an accurate "transition standard" for the calibrated flow rate. CAUTION: The pressure drop in the calibration flow meter should be minimal and should not exceed 0.05 in. Hg.
	C. Calibrating the Coarse-Particle Flow Rate, Q_c:
1	The coarse particle flow should be calibrated at the Control Module using the 1/8″ NPT bulkhead fitting on the side of the enclosure.
2	Disconnect 1/4 in. O.D. coarse-particle flow tubing and fitting.
3	Connect calibration flowmeter to the 1/8″ NPT female bulkhead pipe fitting.
4	Turn pump on. Turn "FLOW ADJUST" knob above the "COARSE PARTICLE FLOW" rotameter on the instrument panel to the calibrated flow rate 1.67 l/min.
5	Record the reading of the "COARSE PARTICLE FLOW" rotameter. This is a "transfer standard" for the calibrated coarse particle flow.
	D. Calibration of Fine-Particle Flow Rate, Q_f:
	Disconnect 3/8 in. O.D. fine-particle tubing and fitting. Connect calibration flowmeter to the 1/4″ NPT female bulkhead pipe fitting.

(continued)

Table 2.20. (continued).

Step	Procedure
	Turn pump on. Turn the "FLOW ADJUST" knob above "TOTAL FLOW" on the instrument panel to the calibrated flow rate of 15.0 l/min. Record the reading of the "TOTAL FLOW" rotameter on the instrument panel. This is a "transfer standard" for the calibrated fine-particle flow. Note that the reading recorded from the "TOTAL PARTICLE FLOW" is *not* the fine particle flow. The fine particle flow is the difference between the total flow.

2.4.3.10 Routine Maintenance

A routine maintenance has been recommended by the manufacturer [8] in order to assure an accurate performance of the Model 245 sample. Description of the necessary routine maintenance is given as follows:

Cleaning the Sampling Module

The virtual impactor head is disassembled by removing the screws. All parts are sealed with "O" rings. Particulate internal loss deposits accumulate primarily on the outer and inner surfaces of the tip of the collection probe in the virtual impactor head. The collection probe should be inspected periodically for such particulate deposits and cleaned as required. A cleaning schedule of every 3–4 months is typical. The remaining inner surfaces should be cleaned every 6–12 months. Cleaning

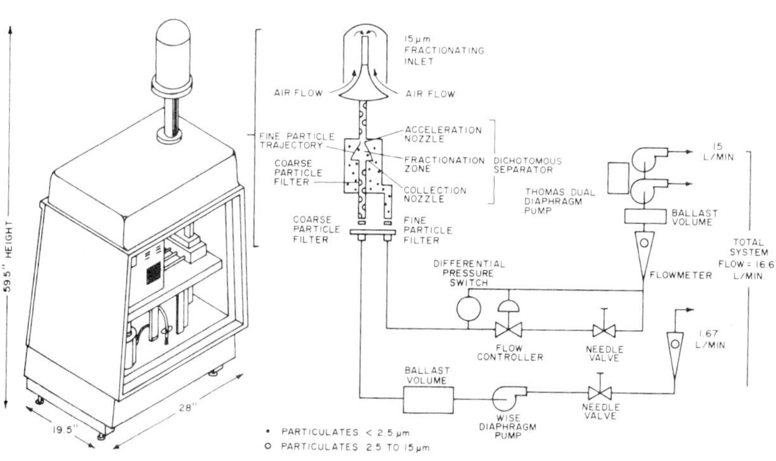

Figure 2.29. Flow diagram of Beckman SAMPLAIR dichotomous sampler used in inhalable particulate network.

should be done with alcohol or water using a camel's hair brush or by washing.

The diametral "O" rings in the aerosol inlet and the flow splitting chamber should be conditioned periodically with vacuum grease.

The bug screen in the aerosol inlet should be cleaned periodically during the summer months. The bug screen is exposed for cleaning by pulling the aerosol inlet off the aerosol inlet tube. A diametral "O" ring in the aerosol inlet acts as the seal.

Control Module

CAUTION: WHENEVER THE FRONT PANEL OF THE MODEL 245 CONTROL MODULE ENCLOSURE IS REMOVED, FIRST UNPLUG THE POWER CORD FROM ITS RECEPTACLE!

Filter Elements—After approximately 12–24 months of sampling the inlet filters for the coarse and fine particle flows may need replacement. In normal operation, air entering these filters is filtered by the two 37 mm membrane filters in the Sampling Module. Hence, inlet filter replacement is infrequent. To replace these filters, first unplug the power cord and then remove the instrument panel of the Control Module by removing the six screws. The filter jar for the coarse-particle flow is the small jar (approximately 1.5″ dia. × 3″ L) in the upper left-hand side of the enclosure. Unscrew the jar, remove the old filter element, and replace with a new element. Tighten the jar very tightly when reinstalling to avoid leaks. The fine-particle filter jar is behind the coarse-particle filter jar. It is approximately 3″ dia. × 5″ L. Filter element replacements are available from Sierra.

Battery Replacement

If AC power has not failed during operation, the dry cell battery for stand-by power for the Digital Controller need be replaced only every 2–3 years. If AC power has failed during operation, as indicated by the dot light on the Digital Controller clock, the battery must be replaced more frequently.

To replace the battery, first unplug the power cord, and then remove the front panel as described above. The battery is located in a bracket on the lower left-hand side of the back wall of the enclosure. The Model 302 BAT Standby Battery (12 VDC) replacement may be used, or equivalent, and is available at the local stores.

Vacuum Pump

The diaphragm of the Model 727CA418 Diaphragm Vacuum Pump is replaced periodically at 1–2 year intervals. First unplug the line voltage, remove the front panel, remove the finned head of the pump by removing

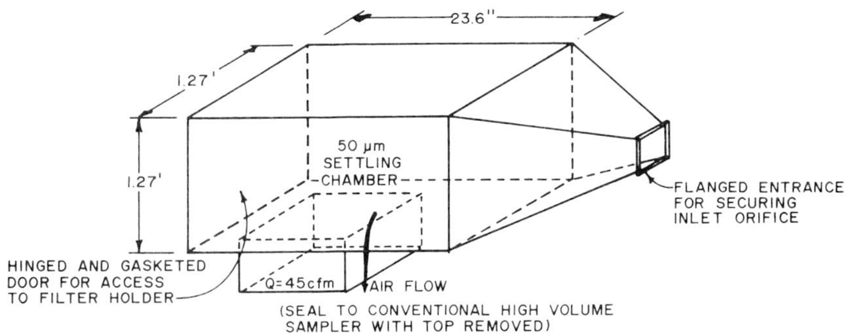

Figure 2.30.a. Sketch of dimensions. Exposure probiling by use of a 50 μm cut diameter isokinetic inlet sampler (this is 5 meters downwind from the source and capable of traversing ±2 meters vertical and horizontal).

the six head screws, and replace the diaphragm. To reassemble, reverse the procedure making sure that the screw clearance cavity in the plate is lined up under intake valve screw heads and that all head screws are tightened down evenly.

Setting Pressure Differential P_1-P_a

The flow controller maintains constant the pressure differential P_1-P_a shown in Figure 2.30. This pressure differential is set in the factory at its optimum value of 4 in. Hg P_1-P_a and can be anywhere in the range of 2 to 6 in. Hg without seriously degrading the performance of the constant flow controller. To check P_1-P_a, first unplug the power cord from its receptacle, remove the front panel, locate the P_1 pressure port located in a tee above the air-line filter with plastic jar located on top of the vacuum pump (exhaust side), remove the 3/8″ tube plug, and insert and tighten a 3/8″ tube connected to a mercury manometer or pressure gauge with the low pressure side open to the atmosphere.

If P_1-P_a is out of the 2 to 6 in. Hg range, readjust by turning the black knob on the flow controller located on the inlet side of the vacuum pump. The flow controller looks like a pressure regulator. The locking screw must first be unloosened. Turning the knob clockwise increases P_1-P_a. After adjusting, be sure to tighten the locking screw and reinsert and tighten the 3/8″ tube plug in the P_1 pressure tap to avoid leaks.

2.4.4 Beckman Automated Dichotomous Sampler

2.4.4.1 Main Features

The Beckman Dichotomous Sampler is used for measuring the mass concentration of IP in ambient environment as it collects simultaneously

Figure 2.30.b. Photo of two isokinetic samplers in loader bucket.

coarse (2.5 to 15 μm) and fine (<2.5 μ) particulates. This sampler works on the same principles as the Sierra Sampler 245 does, with a same total flow rate (16.7 l/min). Therefore the reader can use many of the procedures in Section 2.4.3. The design of the sampler permits manual as well as automatic operation. The filter changing mechanism and seal are problems, however it can be operated manually rather than automatically to circumvent this problem. Figure 2.29 shows the Beckman sampler and the specifications of which are given in Table 2.21. The following sections describe some of the main components of the sampler and operation procedure as documented from the manufacturer's manual [9].

2.5/Special Particulate Sampling Configurations

2.5.1 Introduction

In the investigation of fugitive dust emission (assessment and control), it is important to provide reliable and consistent data on the quantity and characteristics of emissions from the sources under study. Precise analyses can relate good emission data to parameters which characterize source conditions with reliable results. Fugitive dust in-

Table 2.21. Specifications of the Dichotomous Beckman Sampler.

	Specifications
Particulate Size Range:	1 to 15 μm
Inlet Upper Cutoff Point (50%)	15 μm
Virtual Impactor Cut Point (50%): Sharper than ACGIH Criteria	2.5 μm (3.5 μm option)
Virtual Impactor Losses:	<5% for particles between 1 and 15 μm aerodynamic diameter, <10% at cut point
Wind Velocity:	0 to 12 miles (20 km)/hr
Sample Flow Rate:	16.7 liters/min total (1 cubic meter per hour)
Sample Flow Stability:	±5% per week
Timing Accuracy:	±2 min per day
Start Time Selection:	Presettable to any number of days, hours, or minutes up to 365, 23, and 59 respectively
Delay Period Between Samples or Sample Groups:	Presettable to any number of days, hours, or minutes up to 364, 23, and 59
Number of Filters per Sample Group:	Presettable from 1 to 36 filters
Filter Material:	1 μm Teflon standard, 2 μm Teflon optional
Filter Size:	29-mm dia deposit area on 35-mm photoslide type frame
Filter Capacity:	72—36 each for coarse and fine, in 2 trays
Housing:	Weathertight, internally temperature controlled at 33°C
Ambient Operating Temperature Range:	-40° to +122°F (-40° to +50°C)
Operating Humidity Range:	0 to 100% RH
Power Required:	115 Vac, 60 Hz; 200 VA nominal at 25°C with pumps off; 300 VA max at 25°C with pumps on, 350 VA max with heaters and pumps on
Power Failure:	Battery pack maintains timekeeping function for 30 min should power fail; time of power failure is printed.
Power Cord:	8 ft (2.4 m) long, U-ground plug
Dimensions:	40 in. H × 26 in. W × 19-1/2 in. D (102 × 66 × 50 cm); sample inlet protrusion from top requires additional 20 in. (51 cm) height
Instrument Weight:	174 lb (80 kg)
Shipping Weight:	220 lb (100 kg)

vestigations must be carefully chosen relative to the sampling locations, type of samplers, and number of samplers in a given site.

In Section 2.2, description of particulate sampling equipment which are currently used in most IP studies has been presented along with all other necessary operating instructions. These equipment are used individually for the determination of particulate data at the sampling location. However, there exist some investigations that require special sampling methods in order to obtain precise data, such as measurements of source strength or directional measurements of particulate concentrations. The following section describes special sampling methodologies that can be employed in fugitive dust studies.

2.5.2 Isokinetic Directional Measurements Using the High-Volume Samplers

For the measurement of an individual particulate source strength, the two principal difficulties that are usually encountered are the interference of emissions from different sources that may exist in the immediate vicinity, and the change in the wind direction and speed during the sampling period. It is possible to overcome the first problem by sampling the emission source at a short distance directly downwind, so that the background particulate material will be minimal. In addition, correction for any background concentration may be made by simultaneous measurement of the particulate concentration in the upwind direction. Serious errors in the measured concentration are expected if changes in wind speed and direction take place during sampling. It is, therefore, necessary to maintain the sampler intake into the wind as well as maintaining the isokinetic conditions. This can be accomplished using a high volume sampler adapted with special housing provided with an inlet orifice that can be varied depending on the wind speed. This modification to the high-volume sampler has been developed by the authors for field studies of fugitive dust emissions. The shelter roof of the high-volume sampler is replaced by the isokinetic inlet housing. Figure 2.30 shows this assembly. These are placed five meters downwind from the source $\pm$ 2 meters from the source center line (if only one or two are available they should be capable of traversing the $\pm$ 2 meters during sequential, duplicate runs). The enlargement section acts as a diffuser that decelerates the inlet gas flow in the housing. The residence time of the sample in the inlet housing is adequate for the terminal settling of particles larger than 50 μm to be achieved in the duct before being collected on the sampler's filter for normal limestone type dust.

Sets of circular or square orifices with different diameters are prepared so that each orifice can be fixed to the inlet duct for a given range of wind speed in order to maintain the isokinetic conditions.

Figure 2.31 shows the wind speed and the associated orifice size to be used for standard high volume flows of 45 cfm. Before sampling, the prevailing wind speed is to be measured and the suitable orifice can be selected and fixed. If the wind speed changes the sampling can be stopped to replace the orifice as this takes only a few seconds. The high-volume sample can also be moved whenever possible so the orientation of the inlet duct axis coincides with the wind direction during the whole sampling period. This modification to the high-volume sampler provides a size selective inlet that maintains isokinetic sampling in a specified direction.

2.5.3 Exposure Profiling Technique

Qualification of dust emission from potential sources is usually considered as one of the main inputs in fugitive dust studies. It is also a critical part of the study, because reliable emission factors depend upon the accurate measurement of emission rate. Dust is emitted from the source in the form of a plume with a given profile and dimension. The

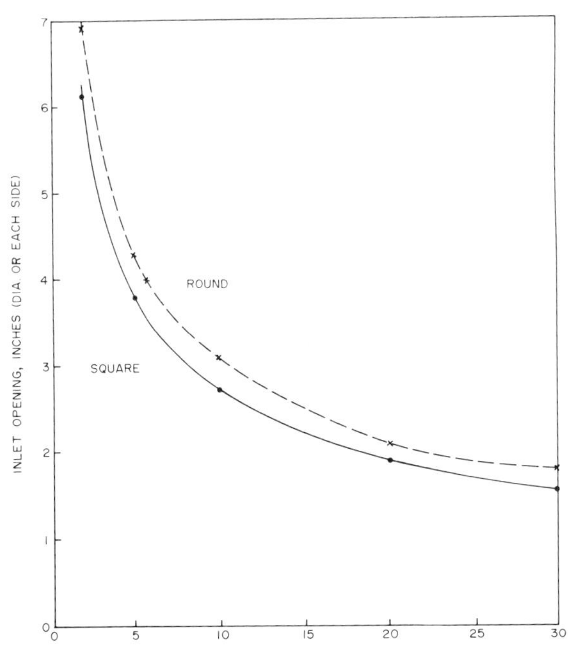

Figure 2.31. Inlet orifice sizes required by isokinetic samplers at various wind speeds (sampler rate is 45 cfm).

latter depend on the size of the source, rate of emission, and the meteorological parameters.

Emission from a source can be quantified by simultaneous isokinetic sampling at different points through the cross-section area of the plume very close to the source to determine the concentrations at these points. The overall plume emission rate may then be determined (for details see Section 2.7.2.2). The larger the number of simultaneous sampling points, the more accurate estimation of the plume emission.

Profilers can be constructed in different arrangements to suit the physical situation of the source and the sampling strategy. Several basic arrangements can be employed in the exposure profiling technique. The first consists of a portable tower 4 to 6 m height supporting an array of four isokinetic high volume filtration devices for sampling total particulate material less than 50 μm attached to a mobile support tower. Each sampler accommodates an 8-in. × 10-in. glass fiber filter. The sampling intake is 2 in. × 2 in. and this increases the allowable wind speed maximum for isokinetic sampling to 20 mph. A flexible hose connects each sampler to a suction manifold. Each leg of the manifold is fitted with a calibrated orifice (connected to an inclined manometer) and a butterfly valve for flow control.

This arrangement is applicable to sources where the ground based profiler tower can be located vertically across the plume and where the distance from the source to the profiling tower can remain fixed at about five meters. Example of such sources are blasting, haul track, vehicles, running on unpaved roads, and agricultural tilling.

The components of the second arrangement of the exposure profiler are typically similar to those utilized in the first type with a grid of six samplers arranged in horizontal and vertical columns, mounted also on a mobile tower. This arrangement is suitable for testing sources like material loading from stockpiles, crushing, or—in general—sources at higher elevation from the ground.

This second arrangement is operated as follows. The sampling intake of each head is pointed into the wind and the sampling velocity is adjusted individually to match the mean wind speed at the location of each sampling head. The wind speed is measured prior to the test at two vertical positions and the vertical wind speed profile is determined by assuming a logarithmic distribution. The result is used to adjust the sampling velocity of each sampling head. During the sampling period, the wind speed is also monitored by recording anemometers at two heights, the sampling velocity can be adjusted whenever necessary to maintain the isokinetic conditions. It has been shown [10] that the logarithmic distribution describes surface winds under neutral atmospheric stability. For short vertical distances (about profiler's height), this distribution holds as a good approximation for other stability classes.

2.5.3.1 Analysis of the Exposure Profiler Results

The exposure profiling technique is usually adopted to estimate the rate of emission from particulate sources. For this purpose, the results can be processed as follows:

Concentration of particulate material

First the weight of particulate collected on the filter, W, is determined by subtracting the weight of filter before sampling from the final weight.

The particulate concentration in $\mu g/m^3$ is then:

$$C = 3.53 \times 10^4 \, W/Q_s t$$

where

Q = sampler flow rate in ACFM
t = sampling period, min.

It is convenient to express the concentration at standard conditions (25 °C and 29.92 in. (760 mm) Hg).

Calculation of the isokinetic flow ratio

In sampling TSP, it is necessary to calculate the isokinetic flow for correcting the particulate concentration in the different size ranges. A pilot tube is used to measure the center line wind speed in the sampling duet, at the point where the ampling probe is installed.

The isokinetic flow ratio R_{iso} is calculated as:

$$R_{iso} = Q_s/U_s \, A_i$$

where the U_s = wind speed approaching the probe ft/min
A_i = intake area of the sampler, ft^2

Determination of the particle size distribution

The profiler provides only the total particulate, TP, concentration. However, cascade impactors may be operated much closer to isokinetic sampling conditions than the exposure profile heads, and detailed particulate size distribution can be obtained.

Calculation of particulate exposure and integration of profiles

The particulate exposure, e (mg/cm^2) at any isokinetic directional sampler is determined as follows:

$$e = W/A_i = 3.05 \times 10^{-8} C_n U_s t$$

where

W = net particulate weight collected, mg.
A_i = sampler intake area, cm^2
C_n = net particulate concentration, $\mu g/m^3$
U_s = approaching wind speed ft/m
t = sampling period, min

The particulate exposure, e, for any given particle size range can be obtained by using the above equation.

The total emission is estimated by integrating the exposure profile at the location of samplers. This can be performed as follows:

a. For sources like blasting, haul truck and vehicles running on unpaved roads where the ground-based profiler tower is used, the integrated profile is given by:

$$W = \int_{0}^{L} W/A \, dy$$

where:

W = the dust catch by exposure sampler (after subtraction of background contribution, g)
L = the effective extent of plume above the ground, m
Y = vertical distance, m
A = intake area of exposure sampler, m^2

b. For elevated sources like material loading from stock piles, crushing . . . etc. in which the second type of profiler is used, the integrated profile is given by

$$W = \int_{-L_{H_1}}^{L_{H_2}} \int_{L_{V_1}}^{L_{V_2}} W/A \, dxdy$$

where:

L_{H_1} and L_{H_2} = the effective extent of plume in the horizontal direction, m
L_{V_1} and L_{V_2} = the effective extent of plume in the vertical direction, M
x and y = horizontal and vertical distances, m.

Determination of particulate emission factor

The emission factor can be expressed in lb or mg of emission per unit of production process. For example if the study is concerned with material loading process, i.e. the unit production process is ton/min. Therefore, the emission factor is:

$$EF = \frac{E}{\text{Process Rate}} = \frac{\text{mg/min}}{\text{ton/min}} = \frac{\text{mg}}{\text{ton}}$$

or it can be converted to any other units of mass using appropriate conversion factors.

2.5.4 Wind Tunnel Method

For precise study of wind erosion process with minimal interference from background sources, the wind tunnel method may be adopted. The method is suitable for the measurement of dust emissions generated by wind erosion of exposed areas and storage piles. The wind tunnel is portable and consists of (see Figure 2.32) an inlet section, a test section, and an outlet diffuser. The test section of the tunnel is an open-floored 1 × 1 foot cross section about 8 ft. long. However, these dimensions may be modified [11] to suit the physical situation of the individual sources under investigation. Air is drawn through the tunnel at controlled velocities. The exit air stream from the test section passes through a circular duct of about 20″ ID; fitted with an isokinetic probe at the downstream end. For source sampling, the open-floored test section is placed directly over the surface to be tested. An airtight seal must be maintained along the sides of the tunnel by rubber flaps attached to the bottom edges of the tunnel sides. The rubber flaps may then be covered

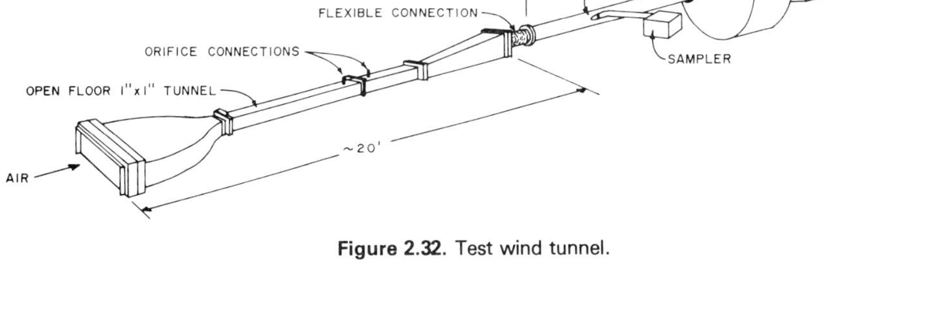

Figure 2.32. Test wind tunnel.

with material from areas adjacent to the test surface to eliminate air infiltration. The local wind speed is measured at the standard 10 m height. The air flow in the wind tunnel is then adjusted to predetermined speed ranges so that it is related to the measured wind speed by means of a logarithmic profile. Tunnel wind speed is measured by a pitot tube at the downstream end of the test section. To reduce the dust levels in the wind tunnel air intake stream, the test may be conducted only when ambient winds are well below the threshold velocity for erosion of the exposed material. The isokinetic sampling probe is connected to a high-volume sampler provided with a cyclone pre-collector and cascade impactor. The sampling train is operated at 15.25 CFM and the size distribution of dust for 30 μm and less (depending on the cut size of the cyclone) is determined from the cascade impactor results. The rate of dust deposition in the test section cannot be measured directly during the test run.

2.5.4.1 Analysis of the Wind Tunnel Results

The wind tunnel data are used to determine the emission rate generated by wind erosion process. The method of calculating the particulate concentration, size distribution, and isokinetic flow ratio are identical to those used for the exposure profiling (Sec. 2.5.3.1). The rest of the calculations are as follows.

Determination of tunnel volume flow rate

The velocity profile along the vertical direction in the tunnel is determined by means of a transversed pitot tube, water manometer, and the use of the following equation:

$$U_y = 6.51\ H_y\ T/P$$

where

U_y = the wind speed at a vertical distance y, m/s
H_y = manometer reading in a vertical distance y, inch of water
T = tunnel air temperature, K
P = tunnel air pressure, inch Hg.

The velocity profile can be integrated over the cross-sectional area of the wind tunnel to determine the volumetric flow rate through the tunnel.

Calculation of particulate emission rate

Using the previously calculated results and the operating data, the emission rate, E ($g/m^2 \cdot s$), due to wind erosion process can be estimated by:

$$E = Cn\ Qt/Ae$$

where

Cn = net particulate concentration, g/m^3
Qt = tunnel flow rate, m^3/s
Ae = tunnel cross section area, m^2

2.5.5 Quasi-Stack Method

This method involves capturing the entire emission streams with enclosures or hoods and applying conventional sources sampling techniques to the confined flow. Accurate result is usually obtained by applying this method. It is not an easy matter to employ such a method unless the physical configuration of the source is suitable for perfect enclosure of its exit.

The enclosure is usually divided into a number of equal areas and exposure profiler samplers provided with flow controllers are placed in the center of each area. Before sampling the wind speed is measured at the center of each area by using a precise velocity meter to determine the wind profile within the flow areas. The wind profile data are used to adjust the sampler flow rates to maintain the isokinetic conditions. The flow rates are checked and adjusted if necessary every 2 to 3 minutes.

The Quasi-stack technique is employed only for sampling a point source, such as stack discharge, drilling, etc. This method has been employed [11] successfully for measurement of the emission rates for drilling (over burden) operation and Figure 2.33 shows the typical quasi-stack arrangement used in this study.

2.6/*Meteorological Monitoring*

2.6.1 Introduction

As any other air pollution problem, the particulate pollution has three consequences, source emission, atmospheric dispersion, and deposition on the receptor. Having been released from a source, fugitive dust is transported and dispersed into the atmosphere. The transport and dispersion are mainly dependent on meteorology. Successful dispersion models are those based on accurately measured and analyzed meteorological data. Turner [12] has pointed out that errors in meteorological data will greatly affect concentration computations. The most important meteorological parameters affecting the transport and dispersion of dust (or any pollutant) are wind speed, wind direction, and consequently the atmospheric stability. Ambient temperature is needed for

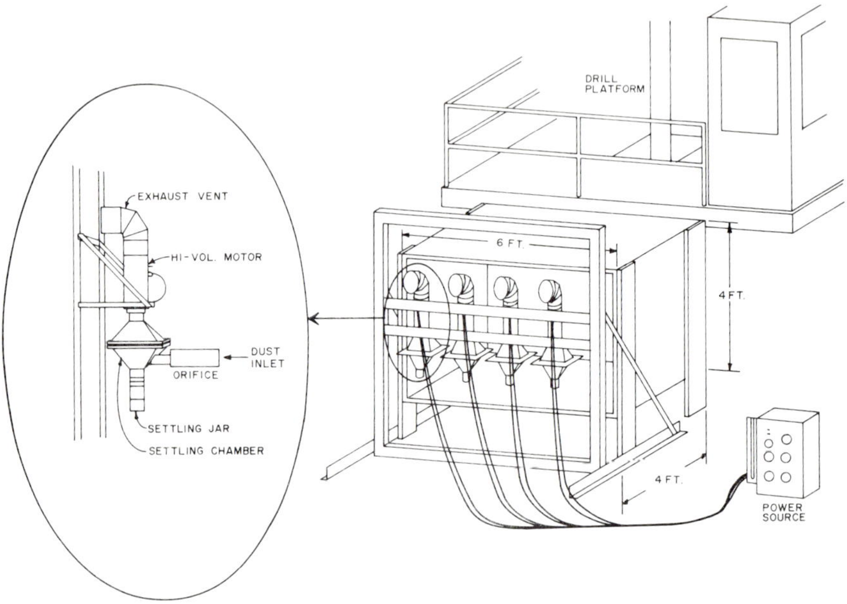

Figure 2.33. Quasi-stack sampling with temporary enclosure for drill sampling.

the analysis of the sampler's data, and the relative humidity is also necessary to interpret the result, especially when the dust concentration is extremely high in the presence of high relative humidity as discussed in Section 1.2. The following sections present the types of meteorological equipment required for measuring and recording the necessary data.

2.6.2 Wind Speed

Surface wind speed is usually measured by the rotating anemometers that provide the necessary data for most air pollution analysis. Of these anemometers, there exist different types such as the windmill or propeller (Figure 2.34) and cup anemometers (Figure 2.35). The propeller anemometers consist of circular array of not less than three blades mounted on one end of a horizontal shaft and on the other end a wind vane is attached. The blades respond to the wind forces and rotate freely. The horizontal shaft attached to a vertical shaft which is mounted on bearings that enable the latter to rotate freely as the varying wind direction acts on the vane. Such a rotation keeps the blades always heading into the wind. The blades rotate a dc generator, the output voltage of which is directly proportional to the wind speed. This type of anemometer is also able to measure wind direction.

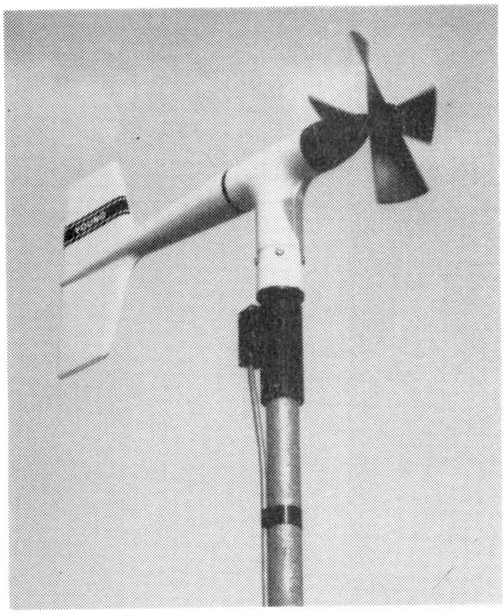

Figure 2.34. Windmill (propeller) anemometer, Model 05102 wind monitor (Courtesy R. M. Young Co.).

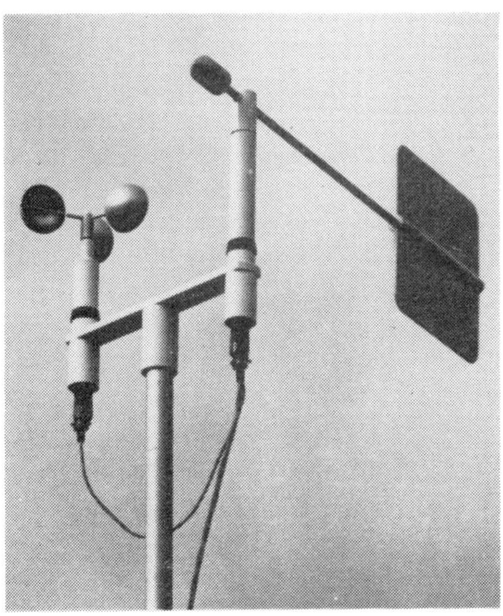

Figure 2.35. Microvane and 3-cup anemometer (Courtesy R. M. Young Co.).

Table 2.22. Distance Constant of Some Anemometers.

Manufacturer	Anemometer Model	Distance Constant, m
R. M. Young Co.	Gill, 3-cup	0.73
Climet Instrument, Inc.	3-cup WS-011	1.5
	4-blade prop. 018-10-39	0.9
Meteorology Research Inc.	Velocity Vane 1057	1.7 to 1.8
	Mechanical Weather Station 1071-1075	5.5
	Vector Vane 1053 III	1.0
Belfort Instrument Co.	Aerovane (3-blade prop) 120	4.6
	3-cup 1250B	7.6
	3-cup Type M	6.4

The rotating cup anemometer (Figure 2.35) consists of three lightweight plastic or aluminum cups 2 to 6 cm in diameter and these are fixed on a cup wheel 5 to 10 cm in diameter from the center of rotation to the center of the cups. The cups are attached by horizontal arms, 120° apart, to a vertical shaft mounted in a bearing and rotates freely. The rate of cup rotation is proportional to the wind speed.

Before selecting the type of anemometer to be employed for a given study, it is necessary to determine the value of lowest wind speed required to be detected by the instrument. The lowest measurable wind speed is called the "sensitivity" of the anemometer. For example, in studies where measurements of air turbulence is required, anemometers with high sensitivity are required. On the other hand, if average wind speeds only are required, anemometers with lower sensitivity may be used. The sensitivity depends mainly on two factors, the mechanical function and the moment of inertia of the rotating vanes of the cup. The measure of sensitivity is the "distance constant." This characteristic value always appears in the instrument's specification. The distance constant, in meters, is defined as the distance the wind passes the anemometer before it responds to a 63 percent change in wind speed. The smaller the distance constant the more sensitive the anemometer. Table 2.22 gives examples of anemometers along with the distance constants [13].

The signal output from the anemometer can be either mechanical or electrical. For example, if the data recording device is located very close to the anemometer, then a mechanical linkage can be used, otherwise an electrical signal is (usually) used as a means of actuating a pen running on a strip chart recorder. The strip chart recorder shows the scale of wind speed versus the time of the day. Recording time for most commercial anemometers is 31 days, and some are over 60 days.

Table 2.23. Damping Ratio and Gust Wavelength for Various Wind Vanes.

Manufacturer	Vane Model	Damping Rate	Gust Wavelength (m)
R. M. Young	Gill Aneometer Bivane	0.60	4.4
Climet Instrument Inc.	Wind vane WD-012-10	0.4	1.0
	Bivane 018-10,-30	0.6	1.0
Meteorological Research Inc.	Wind vane 1022	0.4	1.13
	Vector vane 1053	0.6	1.0
	Mechanical Weather Station 1071-1075	0.5–0.6	6.8
Belfort Instrument Co.	Splayed vane Type M	0.20	9.5
	Aerovane Type L & N	0.30	10.4

2.6.3 Wind Direction

Different types of instruments may be used for measuring the wind direction. The most common types used are the flat plate vane, multiple element vane (Figure 2.35) and airfoil vane. All of these work on the same principle. The vane is mounted at one end of a horizontal rod. The rod is fastened to a vertical shaft mounted in a low torque bearing which moves freely. There are various methods for detecting the wind direction such as:

Multiple-position switch: the vertical shaft has a single contact and four or eight separate stationary contacts positioned in a measurable direction. As the single contact on the shaft makes connection, a signal indicates the wind direction.

Position motors: Two motors are used, one position (transmitting) motor is connected to the wind vane, the other (receiving) can be located in a console or in a recorder with an electronic cable connecting the two motors. The motors are supplied with single-phase AC and any rotation imposed on the first motor (due to wind) is followed by the other motor indicating the wind direction.

Wire potentiometer: This commonly used, movable contact potentiometer (variable resistor) is connected to the vane's vertical shaft. As the shaft rotates, the vertical resistance between the two input terminal changes. This in turn produces the variations on the meter or recorder.

The instrument is initially installed so that the observed reading on the meter or recorder indicates the wind direction in degrees from North, i.e. North = 0° or 360°; East = 90°; South = 180°; and West = 270°.

Table 2.24. Summary of Sample Handling and Analysis.

Sampler	Sample Media	Handling	Analysis	Instrument	Calculation
Cascade Impactor	Foils or fiber filters	In the impactor unit after blocking off the sides	weigh; (option: Chemical analysis)	Analytical balance (sensitivity 0.1 mg)	size distribution
High-volume	Glass fiber filters	Placed in individual envelopes (may be folded)	weigh	Analytical balance (sensitivity 0.1 mg)	mass concentration
Dichotomous Sierra & Beckman	Teflon or fiber glass filters	Field: Transport and store the holders in protective containers. In the Laboratory: The filters are kept in individual sealed petri-dishes (*must not be inverted*)	Weigh	Micro-balance (Sensitivity 1 μg)	mass concentration
Exposure Profiler	Glass fiber filters	Filters are handled as those of High Volume. The interior surfaces of profiler probes and cyclone precollectors are rinsed with distilled water into containers with settling chamber catches. Then completely dried.	weigh	Analytical balance (sensitivity 0.1 mg)	mass concentration
Quasi-stack & wind tunnel	Glass fiber filter	Same as High Volume	weigh	Analytical balance	mass concentration
Use Raw Material (or aggregates)		Not less than five samples are collected in airtight plastic bags	weigh wet; Dry then reweigh	Analytical balance sensitivity 0.1 mg)	moisture content
Use Raw Material		Not less than five samples are collected and completely dried	mechanical sieving	200-mesh screen	silt content
Dust fall	Jars, petri dishes or foils	Jars and Petri dishes are covered, sealed, and transported to the laboratory foils placed in sealed containers.	Dry then weigh	Analytical balance	Rate of dust deposit/unit area

Unlike the wind speed instruments, the sensitivity of the wind direction vanes is given by two parameters:

1. The starting threshold-minimum wind speed required to move the vane.
2. The resistance constant (natural gust wavelength dependent on the mechanical resistance, the moment of inertia, and the design of the instrument)—defined as the length of the wind trajectory necessary to complete an oscillation of the vane.

For describing the response characteristics of the wind vane, a parameter called damping ratio is used to measure the mechanical resistance to the movement of the vane. The lower the value of the damping ratio, the more faithful the wind vane response. Table 2.23 gives examples of damping ratios and gust wavelength for some wind vanes [13].

2.7/Sample Handling

For fugitive dust investigation, samples are usually collected to determine particle size distribution, concentration, and composition including moisture and silt content. Following sampling, the operator must assure the completeness of field data as described in Chapter Three before sending the samples for processing in the laboratory. These samples all required slightly different handling procedures. However, the principal objectives are to minimize sample losses and to prevent contamination and changing of the physical or chemical constitutions. The sample handling, method of analysis, and instruments used in most fugitive dust investigations are summarized in Table 2.24.

References

1. Mercer, T. T., Tillery, M. I., and Newton, G. J., "A Multi-Stage Low Flow Rate Cascade Impactor," *Aerosol Sci., 1,* 9 (1970).
2. *Methods of Air Sampling.* Morris Katz, ed. American Public Health Association (1972).
3. Whitby, K. T., Lundgren, D. A., and Peterson, C. M., "Homogeneous Aerosol Generation," *Intern. J. Air Water Poll., 9,* 263 (1965).
4. Berglund, R. N. and Liu, B. Y., "Generation of Monodisperse Aerosol Standard," *Envi. Sci. & Tech., 7,* 147 (1973).
5. Title 40: Protection of Environment; Part 50: National Primary and Secondary Ambient Air Quality Standard, *Federal Register, 36* (84), 8186 (April 30, 1971).
6. Futune, S. F.G., "Factor Affecting the Precision of High Volume Sampling," M.S. Thesis, University of Florida, Gainesville, Florida (April 1964).
7. Inhalable Particulate Network Operations and Quality Assurance Manual. Office of R&D, U.S. EPA (May 1980).

8. Sierra 245 Instruction Manual, Sierra Instruments, Inc., P.O. Box 909, Carmel Valley, CA.
9. Beckman Automated Dichotomous Sampling System Operating and Service Instructions. FM 675542–301, Beckman Instrument, Inc., 2500 Harbor Blvd., Fullerton, CA 92634 (Feb. 1979),
10. Cowherd, C., Jr., et al., "Department of Emission Factors for Fugitive Dust Sources," Final Report, Midwest Research Institute for U.S. Environmental Protection Agency. Publication No. EPA-450/3-74-037, NTIS No. PB 23826 2/AS (June 1974).
11. Axetell, K., Jr. and Cowherd, C., Jr., "Improved Emission Factors for Fugitive Dust from Western Surface Coal Mining Sources," Vol. I—Sampling Methodology and Test Results. U.S. EPA Contract No. 68-03-2924 (July 1981).
12. Turner, D. B., "Atmospheric Dispersion Modeling. A Critical Review," *JAPCA, 29* (5), 502–519 (May 1979).
13. Noll, K. E. and Miller, T. L. *Air Monitoring Survey Design.* Ann Arbor Science Publishers (1977).

3/Data Processing

3.1/Data Analysis

3.1.1 Determination of Particle Size Distribution

Results of the cascade impactors (Andersen, RAC or any others) are used to determine the particulate size distribution. Hesketh [1] has described a simple method that utilizes the cascade impactors result and determines the particulate log-probability size distribution. The method assumes that half the mass on any stage plus all mass on smaller stages is less than the stage cut diameter (dsa). The authors used this technique for determining the size distribution of the limestone particulates which were sampled by an RAC impactor unit. The typical results are presented in Table 3.1 and the cumulative percent by mass less than d_{50} is plotted against particulate diameter on a log-probability ordinate as shown in Figure 3.1.

As shown in Figure 3.1, the plots result in two straight intersecting lines. This indicates that the limestone particulates consist of more than one type material. This fact has been realized by optical microscope examination for the controlled particles where particles with different colors were seen.

3.1.2 Source Strength

The source strength is the rate of dust emission from a source, e.g. g/min or mass per unit of material. This can be determined by quantifying the amount of dust immediately downwind of the source by means of simultaneous multipoint sampling over the effective cross section of the fugitive emission plume while obtaining process rate weights at the same time. Exposure profiler, wind tunnel, and quasi-stock techniques provide reasonably accurate data for calculating the source strength.

The calculation of the source strength is better explained through a working example using typical test data which were obtained by the

Table 3.1. Limestone Particulate Size Data Obtained by RAC Cascade Impactor.

Stage No.	Stage d_{50} (μm)	Stage Mass collected, g	Mass % on Stage	Cumulative % by Mass less than d_{50}
1	3.00	0.0099	42.49	78.76
2	2.20	0.0077	33.05	40.99
3	1.50	0.0019	8.15	20.39
4	0.94	0.0008	3.43	14.60
5	0.51	0.0025	10.73	7.52
Filter		0.0005	2.15	
Total		0.0233		

authors during the investigation of fugitive dust from a tertiary crusher in a limestone quarry. Two directional isokinetic high-volume samplers (Section 2.5.2) were used simultaneously for sampling over the effective cross section of the plume. The axes of the high volume inlet sections were maintained in the downwind direction during the sampling period.

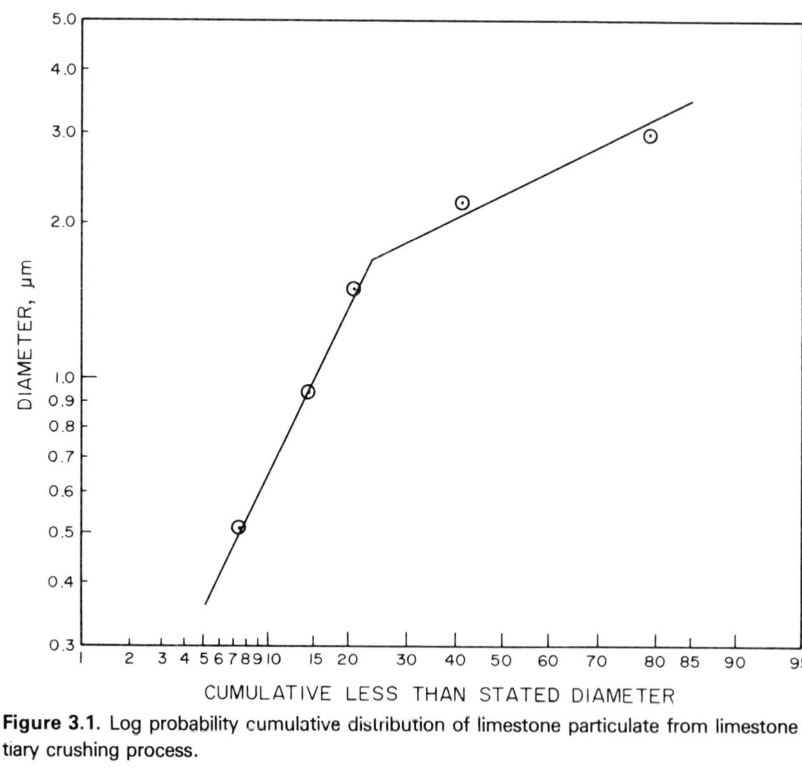

Figure 3.1. Log probability cumulative distribution of limestone particulate from limestone tertiary crushing process.

Test Data

Source: Limestone Tertiary crusher
Samplers: Isokinetic-Directional High Volume Sampler
Distance between sampler and the source = 15 ft.
Plume size at sampling location (by observation) = 24 × 24 ft.
Production (crushing) rate = 867 ton/hr
Meteorological data: Clear sky
  Ambient temperature = 70 °F (21 °C)
  Prevailing Wind Speed = 7 mph
  Relative Humidity = 80%
Sampling period = 3 min.

Figure 3.2 shows the cross-section of the plume approximated by a square and the location of sampling. The rate of dust emission per unit area, q is calculated in g/ft²h as:

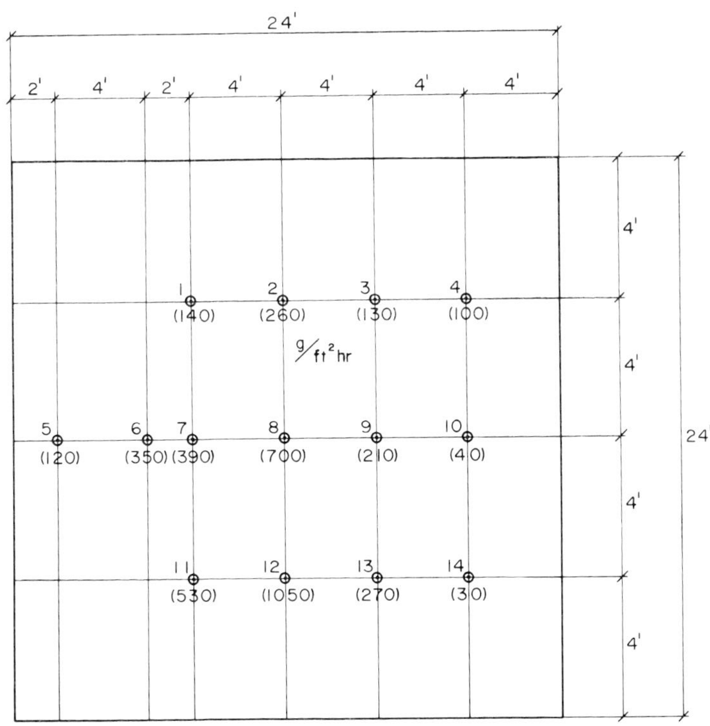

Figure 3.2. Cross section of dust plume 15 ft downwind from source. Typical measured rate of emissions at the sampling points are shown in parentheses in g/(ft² hr) for the 14 numbered sample locations.

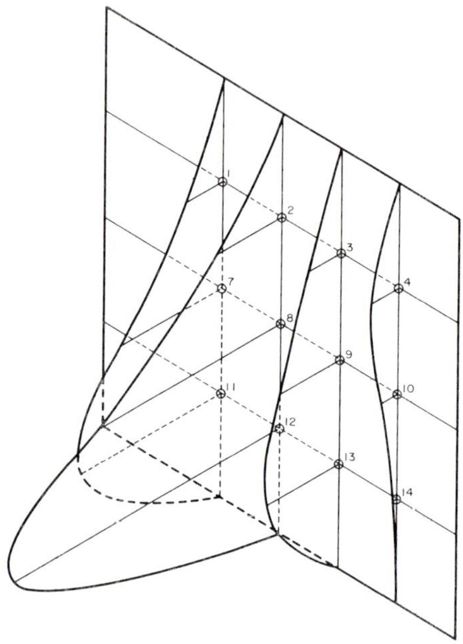

Figure 3.3.a. Measured profile of emission rate per unit area in the downwind vertical direction over the plume cross section for the 14 sampling locations.

$$q = w/at$$

where

w = the dust catch by the sampler, g
a = sampler inlet nozzle, ft^2
t = sampling period, h

Emission profile over the plumes: The rate of emission per unit area is calculated at each sampling location. Figure 3.3.a and Figure 3.3.b shows the profiles of rate of emission per unit area in the vertical and horizontal directions.

Determination of source strength (emission rate): As pointed out in Section 2.5.3.1 that for sources such as the present one, the total emission can be determined by double integration of the rate of emission per unit area over the plume cross-section. The double integration can be performed graphically after simplifying Figures 3.3.a and 3.3.b by replacing them with two single curves representing the geometric average emissions in the two directions. Figures 3.4.a and 3.4.b show the average profiles of emission rate per unit area in the vertical and horizontal directions. Graphical integration can now easily be done by applying the standard method to obtain the rate of emission or the

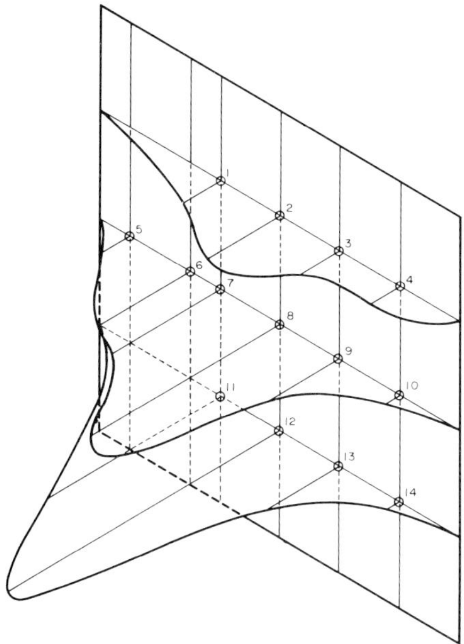

Figure 3.3.b. Measured profile of emission rate per unit area in the downwind horizontal direction over the plume cross section for the 14 sampling locations.

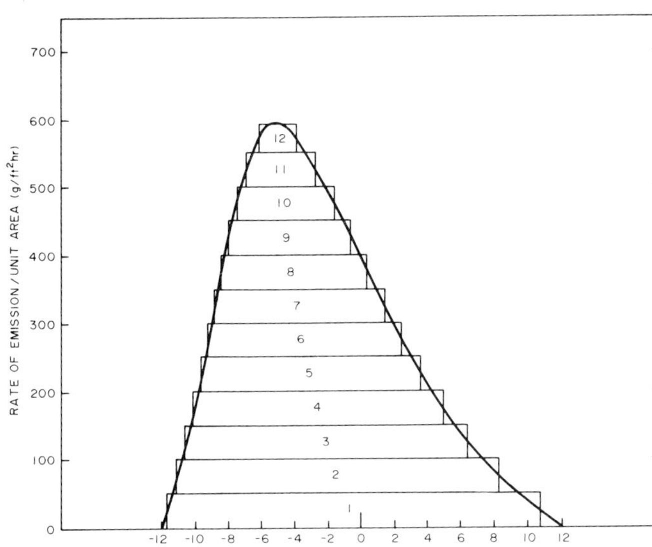

Figure 3.4.a. Average profile of emission rate per unit area in the vertical direction.

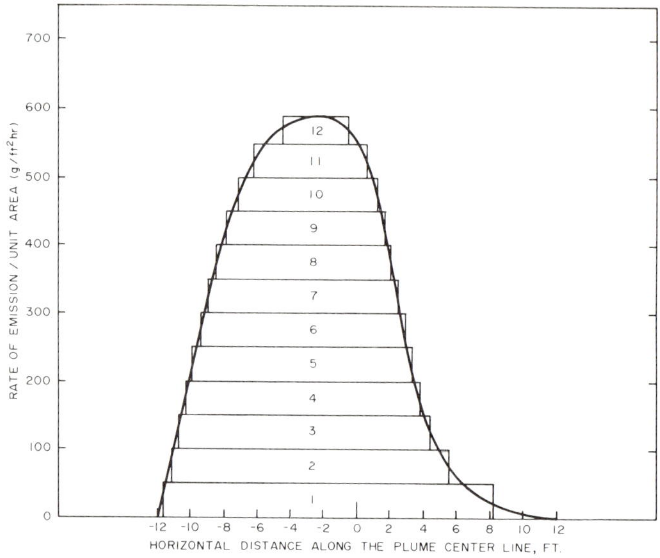

Figure 3.4.b. Average profile of emission rate per unit area in the horizontal direction.

source strength in g/hr. For the present example, the resulting source strength is 98743.5 g/hr.

The steps of a possible method of graphical determination of the emission rate is as follows:

1. Divide the ordinate of Figures 3.4.a and 3.4.b into equal and small increments (each has the unit of g/ft^2 hr). The larger the number of increment, the more accurate would be the result.
2. Draw horizontal lines at each increment so that they divide the area under both curves into number of segments of equal height.
3. Draw two vertical sides for each segment to form a rectangle, the area of which equals the area of the segment. Infinitely small height of segment improves the accuracy of this step.
4. Consider each segment as a three dimensional solid, of a length and width given by the size of each element on Figures 3.4.a and 3.4.b, and the height is the size of increment. The volume of each segment has a dimension of (ft $\times$ ft $\times$ g/ft^2 hr = g/hr).
5. Add up all segment volume to obtain the total rate of emission in g/h.

Table 3.2 gives the details of calculation where Figures 3.4.a and 3.4.b are divided into 12 increments. The emission factor is also given by dividing the rate of emission by the production rate for this example.

Table 3.2. Determination of Source Strength and Emission Factor.

Segment No.	Width (ft)	Length (ft)	Height (g/ft² hr)	Volume (g/hr)
1	22.5	20.0	50	22500
2	19.6	16.8	50	16464
3	17.3	15.5	50	13562.5
4	15.3	14.2	50	10863
5	13.6	13.4	50	9112
6	12.0	12.6	50	7560
7	10.5	11.6	50	6090
8	8.9	10.6	50	4717
9	7.5	9.4	50	3525
10	6.0	8.2	50	2460
11	4.2	6.8	50	1428
12	2.2	4.2	50	462
				98743.5

Emission Rate = 98743.5 g/hr
∴ Emission factor = 98743.5/867 = 113.891 g/ton
= 0.251 lb/ton

3.1.3 Quality Rating of Emission Factors

The emission factor measures the quantity of fugitive dust per unit of source activity. The reliability rating of emission factor has been categorized [2] and updated by the U.S. EPA [3]. The scheme entails the rating of test data quality followed by the rating of the emission factor developed from that test data. The reliability of test data and emission factors ranging from excellent (A) to poor (E). The "A" rating is given to the factor that developed from sound test methodology. Testing techniques which are qualified as a sound methodology are:

1. quasi-stack
2. roof monitor
3. upwind-downwind
4. exposure profiling

Table 3.3 indicates the EPA's revised reliability rating of emission factors.

A common requirement added to the criteria of all ratings cited in Table 3.3 is that, the source category is specific enough to minimize variability within the source category population.

Because of the critical need for reliable emission factors and consistent data, the rating criteria in Table 3.3 must be considered in any fugitive dust study design.

Table 3.3. Reliability Rating of Emission Factor.

Rating	Reliability	Criteria
A	Excellent	'A'—rated test data taken from many randomly chosen facilities in industrial population.
B	Above Average	'A'—rated test data from reasonable number of facilities. It is not clear if the facilities represent a random sample.
C	Average	'A' and 'B'—rated test data from a reasonable number of facilities. It is not clear if the facilities represent a random sample.
D	Below Average	'A' and 'B'—rated test data from a small number of facilities, and the random choice is suspectable.
E	Poor	'C' and 'D'—rated test data, and the random choice is suspectable

3.1.4 Development and Derivation of Emission Factors

Section 3.1.2 explains the calculation steps of the rate of emission through a working example. If the rate of emission (98743 g/hr) is divided by the production rate (867 ton/h), the result will be the emission factor, EF, of 113.9 g/ton (0.25 lb/ton). This factor must be reproduced by means of several tests. The final emission factor for the source under investigation is derived simply as the arithmetic mean of the individual emission factor values.

It is very useful to supplement the fugitive dust investigation with a statistical analysis of the results to produce a predictive equation describing the emission factors in terms of parameters that characterize source and climatic conditions. These parameters may be categorized as follows:

1. Measures of sources activity or energy expended (e.g., the speed and weight of vehicle travelling on an unpaved road).
2. Properties of the material being disturbed.
3. Climatic parameters.

Advantage of the predictive equations is their application to a much wider range of sources than single-valued factors. Hesketh and Cross [4] have reviewed published predictive equations of uncontrolled emission factors for different sources along with the reliability ranks of numeric values of these factors.

The predictive equation of emission factors are usually derived by regression analysis of test data. The outline of regression analysis is given in the following section.

3.1.5 Application of Multiple-Linear Regression (MLR) Analysis

Multiple Linear Regression is a statistics package for analyzing the functional relation between several independent variables (say x_1, x_2, . . . etc.) and a single dependent variable (say y). The technique uses the method of least square to determine a linear prediction equation from a set of simultaneously-obtained data prints for all variables.

The resulting equation is in the form:

$$y = a_0 + a_1 x_1 + a_2 x_2 + \ldots\ldots a_n x_n$$

where a_a to a_n are coefficient which minimize the squared difference between the fitted function and the data. Multiple Linear Regression is usually available as a computer program in the Statistical Package for Social Science (SPSS) [5].

For example, if measurements were made for emission (dependent parameter) rate from a given source along with all variable (independent parameter) that are believed to influence the value of emission, then the (MLR) steps must begin with:

1. Prepare a data array of all monitored independent variables with corresponding emission measurements.
2. Input these data into the MLR program to transfer both independent variables to their natural logarithm. (Most programs include this step.) This step is required in order to obtain an emission factor equation including multiplicative correction factors rather than additive.

As the program runs, the output gives the following:

1. The constant term (a_0) and the individual coefficient of variables a_1 to a_n.
2. Correlation coefficients of all variables. This is for identifying two interdependent variables (two variables that produce the same effect on the dependent variable), one of which should be eliminated from the analysis.
3. The multiple correlation coefficient. The square of the multiple correlation coefficient is the fraction of total variance in the dependent variable that is accounted for by the variables in the equation at that point.
4. Residual coefficient of variability, which is the standard deviation of the dependent parameter (here, the emission rates) by the equation divided by the mean of the predicted emission rates (expressed as a percent). If a variable eliminates some sample variance, it will reduce the standard deviation and hence the relative coefficient of variability.
5. Significance of regression as a whole. This value is calculated from a

number of tests by comparing the variance accounted for by the regression equation to the residual variance. A 0.05 significance level is a 1 in 20 chance of the correlation being due to random occurrence.

6. Significance of individual variable. This is a measure of whether the coefficient "a" is different than 0, or the relationship with the dependent variable is due to random occurrence. Variables that do not meet a pre-specified level (a choice depends on the study characteristics) may be eliminated from the equation.

3.2/*Quality Assurance*

Quality assurance (QA) is a technique of specifying procedures and measures to be utilized to help assure the validity of the data. These steps include all prefield measurements, actual field sampling, sample handling, analytical procedures, and reporting of the data. It is important that these steps be formalized to some degree so that misleading and erroneous information is discarded before it misdirects the study or becomes part of the results. When in doubt, EPA QA procedures can be used as guidelines [6].

Examples of quality objectives/requirements and suggested ranges are:

Continuous Analyzers—calibrate before and after each test; calibrate hourly for tests > 2 hrs. duration; correct data if possible or else discard data if drifts > ±2 percent occur.

Instruments—Precision standard deviation of < ± 10 percent; accuracy deviations of < 5 percent.

Completeness—Applies to 90 percent of all samples.

Blanks—Analyze one blank for every 3 samples.

3.2.1 Sampling Requirements

The field test teams should be given test plans identifying the equipment to be tested, the materials to be tested, the testing cycle (duration and frequency), the method of operation, and the sampling methods to be employed during the test run. It may be advisable to work from a detailed flow schematic in complicated sampling situations. A *bound field notebook* should be used *by the team leader* to enter all records of pertinent facts applicable to each test run. A *separate bound notebook should be used by the sample custodian* to record all the identifying information recovered. This will be cross-checked with the list of samples that are scheduled to be collected during each run. A parametric screening test sequence can be used initially to reduce the number of subsequent sample needs.

The continuous analyzers will be operated according to the design specifications outlined in the appropriate operating manuals. The particulate mass and particulate size instruments will be operated according to standard procedures such as those discussed in Section 2.

The particulate matter collected may be analyzed immediately upon collection for various substances such as carbon, sulfates, nitrates, and metallic cations. Particulate samples may also be stored for possible future analysis including trace species. All cascade impactor type particulate samples should be transferred to clean polystyrene or polyethylene containers of appropriate size. These bottles will be cleaned by procedures specified by EPA for Level 1 sampling containers [1]:

1. Prepassivate all container and cap surfaces that could contact the sample by a one-hour standing contact with 1:1 vol to vol aqueous nitric acid.
2. Three successive washes with the following solvents:
   a. Distilled water
   b. Isopropyl alcohol
   c. Methylene chloride
3. After washing, purge dry in a filtered stress of dry air or dry nitrogen, then inspect for any residue.

Particulate samples should be delivered to the laboratory as soon as possible and no longer than one week from date of collection. At the end of any test period where a significant change in operation occurs, all samples should be submitted at that time for laboratory analysis. Prior to submitting samples to the laboratory for analysis, the proposed analytical methods and quality control procedures to be used in the laboratory should be reviewed.

3.2.2 Sample Custody

A sample custodian should be appointed. This person is assigned responsibility of the samples and for the physical acceptance of each group of samples intended for treatment or analysis. An important quality of the sample custodian is completeness, checking of records and data, and identity of the samples with respect to the test plan.

Each sample should be assigned a number which defines the date and time that the sample was collected and recovered, the sample train and the system from which the sample was taken. This information will be recorded in a bound notebook maintained specifically for the test site. The samples actually recovered from each run should be compared with

Table 3.4. Calibration of Test Equipment.

Item	Calibration Method	Frequency Calibration	Reference Standard Used
Barometer	±2.5 mm from STD	Each field site	Mercury barometer
Thermocouple	3 pt. calib ±1.5%	Each field site	Mercury-in-glass thermometers
Dry gas meter	3 pt. check vs. STD within ±2%	Each field site	Wet test meter
Orifice meter	Positive displacement	Yearly	Primary standard positive displacement
Pitot tube	EPA Method 2 specification	3 field sites or 3 months	Standard pitot tube or NBS type S pitot tube
Analytical balance	±1 mg of Class S weights	Each field site	Class S weights

a list of samples which are scheduled to be recovered to determine completeness of samples. All samples should be accounted for when transferred and a sample tracking report sheet identifying all samples should accompany the samples at all times.

Any unusual events should be logged in the log book and reported to the quality monitor at the time of occurrence or upon completion of the analysis, depending upon the incident severity. On a periodic basis, draft reports of results should be reviewed by the individual assigned as quality monitor.

3.2.3 Calibration

The measurement systems must be calibrated to assure the quality of the sampling system. Some typical equipment, as an example, and calibration notes are listed in Table 3.4. All of these procedures are suggested by the EPA procedures handbook [1].

3.2.4 Analytical Procedures

Section 3.2.1 noted three of the EPA procedures required for quality assurance of cascade impactor samples. In addition, EPA procedures specified by the IERL (Industrial Environmental Research Laboratory) should be applied in the cleaning of glassware and in all aspects of sample recovery. "Level 1" procedures are recommended for this [6].

3.2.5 Validation and Reporting

3.2.5.1 General

Reviews as to completeness of data forms, notebooks, and sample and process information should be performed daily. At the end of each sample run where samples are collected, be sure that all samples are labeled properly and that all samples scheduled to be collected have been collected. The person in charge of sampling will check daily to make certain that all data forms have been completed and necessary calibrations performed and documented. The supervisor should check both the field and sample recovery notebooks weekly to verify completeness, compatibility of units, and data validity.

Various individuals should be assigned responsibility for assuring that specific samples are taken (e.g. raw materials, fuels, products) and recorded. A report format should be devised to assist in this.

3.2.5.2 Internal Checks

True replicate particle samples are not usually possible to obtain with most particulate samples. However, as much as possible, use duplicate test conditions and samples. Periodic use of calibration standards can also be submitted as samples.

3.2.5.3 Audits

Auditing procedures should be the responsibility of the supervisor. The person (or persons) responsible for this should develop a checklist of items or activities to be inspected each week to determine whether the test team is following the established procedures for sample collection and recovery. The results will be discussed each week with the persons sampling to decide whether procedures or activities should be modified. Audits by independent, knowledgeable persons would be invaluable to the quality assurance of the program. Corrective actions should be initiated without delay to maintain integrity of the program.

References

1. Hesketh, H. E. *Air Pollution Control.* Ann Arbor Publishers (1979).
2. "Compilation of Air Pollution Emission Factors," U.S. EPA #AP-42 with supplement (January 1984).
3. "Fugitive Dust Emission Factor Update for AP-42," U.S. EPA Contract No. 68-02-3177 (December 1982).
4. Hesketh, H. E. and Cross, F. L., Jr. *Fugitive Emission and Controls.* Ann Arbor Publishers (1983).
5. Nie, N. H., et al. *Statistical Package for the Social Sciences.* 2nd Edition, New York: McGraw Hill, Inc. (1975).
6. "Quality Assurance Handbook for Air Pollution Measurement Systems. Vol. III—Stationary Source Specific Method," EPA-600/4-77-0276, August 1977 (revised January 4, 1982).

Biographies

Howard E. Hesketh is a Ph.D. chemical engineer and a registered professional engineer. He is a professor at Southern Illinois University at Carbondale, where he instructs courses in engineering economics, thermodynamics, and air pollution control. He has worked extensively with industry, having served as a consultant to a number of industrial facilities as well as to U.S. Federal and local governmental agencies, foreign government agencies, and other organizations. Dr. Hesketh is author of three engineering texts, has co-authored several other books, and is editor of several prominent publications. He is a certified air pollution control engineer by the American Academy of Environmental Engineers. Currently he is a board member for both the Air Pollution Control Association and the American Academy of Environmental Engineers.

Mohammad S. El-Shobokshy is an associate professor at the King Saud University, Riyadh, Saudi Arabia. He has a BSc in Mechanical Engineering from the Assiut University, Egypt, and an M.Eng. and Ph.D. in Mechanical Engineering from Liverpool University, England. His industrial experience includes work in establishing the Aswan High Dam in Egypt and he was a consultant for the Saudi Fund for Development.

Dr. El-Shobokshy is a member of the Environmental Pollution Committee of the Saudi National Center for Science and Technology and a member of the Air Pollution Control Association. He is the organizer of the environmental pollution course for engineers and environmentalists in Riyadh, Saudi Arabia. He is currently conducting research on atmospheric particulate and lead pollution in the city of Riyadh and the effect of fugitive particulates on solar collecting devices. He has several publications in the area of aerosol and particulate transport.

During 1982–1983 he was a visiting professor in the Department of Thermal and Environmental Engineering, Southern Illinois University at Carbondale, Illinois, U.S.A. There he participated in an investigation of fugitive dust emissions from limestone crushing operations.

Index